Springer Series in

ADVANCED MICROELECTRONICS 26

Springer Series in

ADVANCED MICROELECTRONICS

Series Editors: K. Itoh T. Lee T. Sakurai W.M.C. Sansen D. Schmitt-Landsiedel

The Springer Series in Advanced Microelectronics provides systematic information on all the topics relevant for the design, processing, and manufacturing of microelectronic devices. The books, each prepared by leading researchers or engineers in their fields, cover the basic and advanced aspects of topics such as wafer processing, materials, device design, device technologies, circuit design, VLSI implementation, and subsystem technology. The series forms a bridge between physics and engineering and the volumes will appeal to practicing engineers as well as research scientists.

18 **Microcontrollers in Practice**
By I. Susnea and M. Mitescu

19 **Gettering Defects in Semiconductors**
By V.A. Perevoschikov and V.D. Skoupov

20 **Low Power VCO Design in CMOS**
By M. Tiebout

21 **Continuous-Time Sigma-Delta A/D Conversion**
Fundamentals, Performance Limits and Robust Implementations
By M. Ortmanns and F. Gerfers

22 **Detection and Signal Processing**
Technical Realization
By W.J. Witteman

23 **Highly Sensitive Optical Receivers**
By K. Schneider and H.K. Zimmermann

24 **Bonding in Microsystem Technology**
By J.A. Dziuban

25 **Power Management of Digital Circuits in Deep Sub-Micron CMOS Technologies**
By S. Henzler

26 **High-Dynamic-Range (HDR) Vision**
Microelectronics, Image Processing, Computer Graphics
Editor: B. Hoefflinger

Volumes 1–17 are listed at the end of the book.

B. Hoefflinger (Ed.)

High-Dynamic-Range (HDR) Vision

Microelectronics, Image Processing, Computer Graphics

With 172 Figures

Springer

Professor Dr. Bernd Hoefflinger
Director (retired)
Institute for Microelectronics Stuttgart
Leonberger Strasse 5
71063 Sindelfingen, Germany

Series Editors:
Dr. Kiyoo Itoh
Hitachi Ltd., Central Research Laboratory, 1-280 Higashi-Koigakubo
Kokubunji-shi, Tokyo 185-8601, Japan

Professor Thomas Lee
Stanford University, Department of Electrical Engineering, 420 Via Palou Mall, CIS-205
Stanford, CA 94305-4070, USA

Professor Takayasu Sakurai
Center for Collaborative Research, University of Tokyo, 7-22-1 Roppongi
Minato-ku, Tokyo 106-8558, Japan

Professor Willy M. C. Sansen
Katholieke Universiteit Leuven, ESAT-MICAS, Kasteelpark Arenberg 10
3001 Leuven, Belgium

Professor Doris Schmitt-Landsiedel
Technische Universität München, Lehrstuhl für Technische Elektronik
Theresienstrasse 90, Gebäude N3, 80290 München, Germany

ISSN 1437-0387

ISBN-10 3-540-44432-7 Springer Berlin Heidelberg New York
ISBN-13 978-3-540-44432-9 Springer Berlin Heidelberg New York

Library of Congress Control Number: 2006933223

Springer is a part of Springer Science+Business Media.

springer.com

Typesetting by the editor and SPi using Springer LaTeX macro package.
Cover concept by eStudio Calmar Steinen using a background picture from Photo Studio "SONO".
Courtesy of Mr. Yukio Sono, 3-18-4 Uchi-Kanda, Chiyoda-ku, Tokyo
Cover design: eStudio Calmar Steinen

Printed on acid-free paper SPIN: 11574088 57/3100/SPi - 5 4 3 2 1 0

Preface

The high-fidelity vision of our world has been a continuous challenge as our intelligence and skills have evolved. The acquisition and the mapping of the rich and complex content of visual information rank high among our most demanding technical tasks.

Electronic vision took big leaps with image sensors capturing in real time a dynamic range from bright to dark of more than seven orders of magnitude, thus exceeding the spontaneous ability of the human eye by more than a hundred times, and with displays displaying five orders of magnitude in brightness, again manifesting a more than 100-fold improvement of the state of the art.

This book is the account of high-dynamic-range (HDR) real-time vision starting with high-speed HDR photometric video capture and following the video-processing chain with coding and tone mapping to the HDR display.

The power of eye-like photometric video is demonstrated in machine-vision, medical and automotive applications, and it is extended to HDR sub-retinal implants for the vision impaired.

While the book tries to convey the overall picture of HDR real-time vision and specific knowledge of microelectronics and image processing is not required, it provides a quantitative summary of the major issues to allow the assessment of the state of the art and a glimpse at future developments.

Selected experts have been invited to share their know-how and expectations in this rapidly evolving art surrounding the single most powerful one of our senses.

Stuttgart, July 2006 Bernd Hoefflinger

List of Contributors

Dipl.-Ing. (FH) Daniel Brosch
IMS CHIPS
Allmandring 30a
70569 Stuttgart
Germany
brosch@ims-chips.de

Dr. Alexander Dollberg
University of Dortmund, AG Microelektronic
Emil-Figge Str.68
44227 Dortmund
alexander.dollberg@uni-dortmund.de

Dr.-Ing. Michael Goesele
University of Washington
Department of Computer Science and Engineering
Box 352350
Seattle, Washington 98195-2350
USA
goesele@cs.washington.edu

Dipl. Phys. Heinz-Gerd Graf
IMS CHIPS
Allmandring 30a
70569 Stuttgart
Germany
graf@ims-chips.de

Dr. Volker Gengenbach
GEVITEC
Marie-Curie-Straße 9
76275 Ettlingen
Germany
gen@gevitec.de

Dr. Christine Harendt
IMS CHIPS
Allmandring 30a
70569 Stuttgart
Germany
harendt@ims-chips.de

Prof. Dr. Bernd Hoefflinger
Leonberger Strasse 5
71063 Sindelfingen
Germany
bhoefflinger@aol.com

Dr. Klaus-Martin Irion
Karl Strorz GmbH & Co. KG
Mittelstr. 8
78532 Tuttlingen
Germany
K.Irion@karlstorz.de

Prof. Dr. Peter M. Knoll
Robert Bosch GmbH
Postfach 16 61
71226 Leonberg
Germany
peter.knoll@de.bosch.com

Dipl.-Ing. Grzegorz Krawczyk
Max-Planck-Institut für Informatik
Dept. 4: Computer Science
Stuhlsatzenhausenweg 85
66123 Saarbrücken
Germany
krawczyk@mpi-inf.mpg.de

Dr.-Ing. Roman Louban
hema electronic GmbH
Röntgenstrasse 31
73431 Aalen
Germany
r.louban@hema.de

Dipl.-Ing. Rafal Mantiuk
Max-Planck-Institut für Informatik
Dept. 4: Computer Science
Stuhlsatzenhausenweg 85
66123 Saarbrücken
Germany
mantiuk@mpi-inf.mpg.de

Dr.-Ing. habil. Karol Myszkowski
Max-Planck-Institut für Informatik
Dept. 4: Computer Science
Stuhlsatzenhausenweg 85
66123 Saarbrücken
karol@mpi-sb.mpg.de

Dipl.-Ing. Michael Raasch
hema electronic GmbH
Röntgenstrasse 31
73431 Aalen
Germany
m.raasch@hema.de

Dipl.-Ing. Bela Michael Rohrbacher
hema electronic GmbH
Röntgenstrasse 31
73431 Aalen
Germany
m.rohrbacher@hema.de

Dipl.-Ing. Hans-Joerg Schoenherr
KST GmbH – Kamera & System Technik
Kamera & System Technik
Hugo-Küttner-Strasse 1a
01796 Pirna
Germany
hj.schoenherr@kst-dresden.de

Dr. Helge Seetzen
BrightSide Technologies Inc.
1310 Kootenay Street
Vancouver, BC
Canada V5K 4R1
Helge.seetzen@brightsidetech.com

Dipl.-Inf (FH) Verena Schneider, MSc
IMS-CHIPS
Allmandring 30a
70569 Stuttgart
Germany
schneider@ims-chips.de

Dipl.-Ing. Jan-Dirk Schulze Spüntrup
IMS-CHIPS
Allmandring 30a
70569 Stuttgart
Germany
schulz@ims-chips.de

Dipl.-Ing. Markus Strobel
IMS-CHIPS
Allmandring 30a
70569 Stuttgart
Germany
strobel@ims-chips.de

Dipl.-Ing. (FH) Karsten Warkentin
IMS-CHIPS
Allmandring 30a
70569 Stuttgart
Germany
warkentin@ims-chips.de

1

The Eye and High-Dynamic-Range Vision

Bernd Hoefflinger

The dream of electronic vision is to mimic the capabilities of the human eye and possibly to go beyond in certain aspects. The eye and human vision overall are so versatile and powerful that any strategy to realise similar features with electronics and information technology have to be focused on certain aspects of man's powerful sense to collect the most comprehensive information on our physical world.

The single most challenging feature of optical information is the high dynamic range of intensities hitting the same receptor from one instant to another or hitting different receptors in an observed scene at the same time. This range may exceed 8 orders of magnitude (Table 1.1).

This is wider than our eyes can handle instantaneously or even with long-time adaptation. The sensing and acquisition system in our eyes handles over 5 orders of magnitude in real time (Fig. 1.1) and up to 8 orders with long-time adaptation. The curve shown here represents the instantaneous response of a normal eye. The minimum detectable signal moves up and the maximum tolerable because of overflow (blinding) and damage decreases with age so that the dynamic range deteriorates with age. In any event, the eye's dynamic range used to be much superior to any real-time electronic acquisition system until the advent of certain high-dynamic-range (HDR) video sensors and cameras.

It is the cornerstone of this book that real-time HDR video acquisition with a dynamic range of more than 7 orders of magnitude has evolved to a mature technology since it was first demonstrated in 1992 [1]. Fundamentally, the pixels in these sensors mimic the logarithmic compression of the high input range similar to the photoreceptors in our eyes making them "bionic" sensors in many respects like contrast sensitivity, separation of the illuminant and color constancy [2]. Cameras with such sensors have been available since 1996 and a significant amount of research and development along the vision chain including video processing and the challenging art of displaying HDR video to our eyes has been performed.

Table 1.1. Optical intensities in photometric units in some real-world scenes

Condition	Illuminance (lx)
Clear night sky	0.001
Quarter moon	0.01
Full moon	0.1
Late twilight	1
Twilight	10
Heavy overcast	100
Overcast sky	1,000
Full daylight	10,000
Direct sunlight	100,000

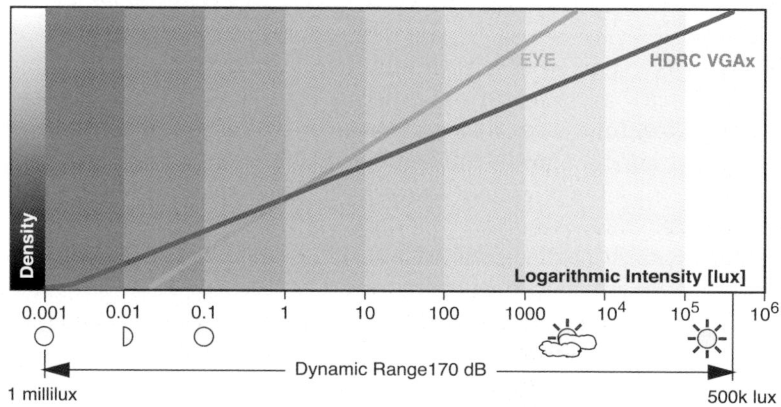

Fig. 1.1. The response, in the following also called the optoelectronic conversion function (OECF), of the human eye and of an HDRC array sensor

As a result, a comprehensive and exemplary treatment of HDR video can be presented, which, as we shall see, is inspired in many ways by the unique capabilities of the human eye and the human visual system (HVS).

The eye-like HDR video acquisition paradigm in this book is evident in Fig. 1.1, where the OECF of a commercial High-Dynamic-Range CMOS (HDRC) sensor is overlayed on the eye's OECF.

What is striking is that the electronic sensor is superior to the eye in:

1. Sensitivity, which means the minimum detectable illuminance, by at least a factor of 10
2. Range, which means the maximum illuminance measurable without distortion, white saturation or damage by another factor of 10, so that the total dynamic range is more than 100× higher than that of the human eye

The unique feature of this sensing technology is HDR photometric video capture in real time of 8 orders of magnitude of illuminance.

This means that over an illuminance range of 100 million to 1 all values measured consecutively on the same pixel and/or simultaneously on different pixels in an array are correct relative to each other and on an absolute scale, if the sensor temperature is kept constant and the sensor is calibrated.

Before we treat other unique features and consequences of this paradigm of real-time photometric HDR video capture, it is useful to put our treatment of HDR vision in perspective relative to the book HDR IMAGING by Reinhard, Ward, Pattanaik and Debevec [3], which we will reference as HDRI in the present book (HDRV).

HDRI is a formidable scientific treatment on the "processing" of high-dynamic-range images acquired with low-dynamic-range cameras basically through multiple exposures. This means HVI is focused on high-quality still images and computer graphics while there are limitations for video, particularly high-speed video in real-world scenes.

The present book HDRV presents the disruptive technology of HDR "acquisition" with eye-like log-compression up front in the electronic photoreceptor (pixel) and its powerful consequences for robust, real-time video processing, transmission and display. Our focus on real-time HDR video capture is demonstrated in Figs. 1.2–1.4.

Fire breather, solar eclipse and welding simply could not be captured without the high-speed, real-time recording offered by the HDRC sensors. However, even with the "natural", eye-like log compression in each pixel, the camera output data for these scenes still needed "tone mapping", which will be presented

Fig. 1.2. A frame taken from the video sequence of a fire breather recorded with an HDRC video camera at 30 frames/s^{-1}

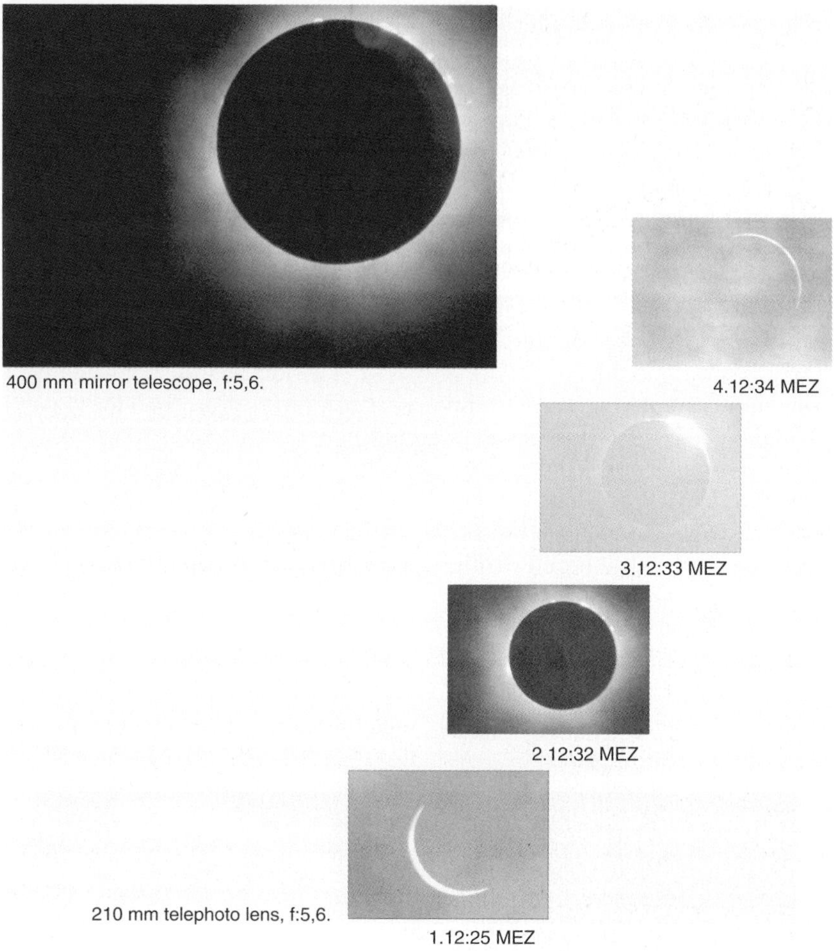

Fig. 1.3. Solar eclipse: a few frames from the video sequence of the solar eclipse on August 11, 1999, in Germany, recorded with a HDRC camera at 30 frames/s^{-1}. The camera was equipped with a 400 mm telescope, f:5.6, and aH6 filter. The partially clouded dark sky was genuinely captured with its shady details because of the low-light contrast sensitivity of the HDRC sensor simultaneously with the extremely bright corona

in later chapters, to offer us an informative print image. It is in areas like tone mapping where all in the HDR community come together to benefit from the general progress in HDR imaging.

However, we return here to HDR eye-like logarithmic video acquisition to identify further tremendous benefits resulting from mimicking the HVS.

Figure 1.5 shows HDR acquisition. The scale in this graph is powers of 2, also called f-stops or octaves. The upper bars show how seven exposures

Fig. 1.4. Welding scence: Single frames from the video sequence of a welding scene recorded simultaneously with a CCD and an HDRC camera in 1995. The HDRC sensor had a resolution of 256 × 128 pixels and it delivered 120 frames/s^{-1}

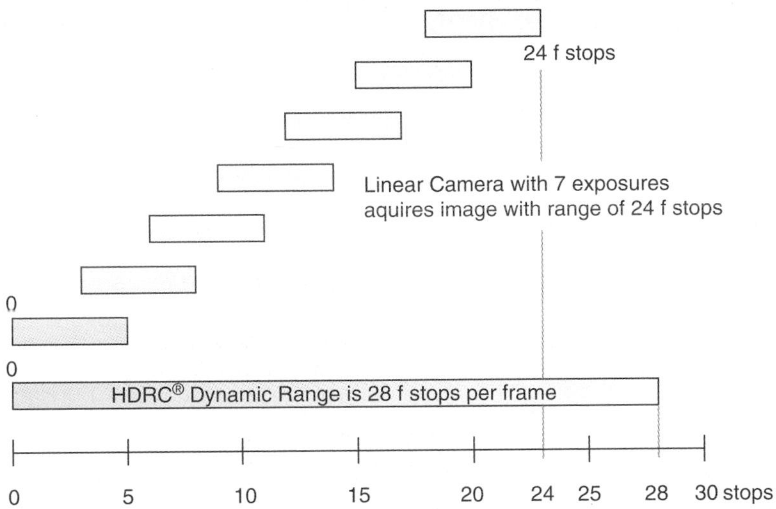

Fig. 1.5. HDR acquisition

with a conventional camera are staggered to obtain an HDR image file with a dynamic range of 23 f-stops, considered to be sufficient for a high-quality image file. Special stitching algorithms are required to fit the 7 files together into one comprehensive file. This art is covered extensively in HDRI Chap. 4 [3]. By comparison, HDRC VGAX sensors have been reported to have an instantaneous range of 8 orders of magnitude or 28 f-stops [2] so that one exposure captures all the information photometrically correct.

The optoelectronic conversion function (OECF) of the HDRC sensor, Fig. 1.1 is not only continuous and monotonic over so many orders of magnitude, it is moreover strictly logarithmic over the largest part. With its extremely high dynamic range, it reduces radically the design and operating problems of cameras:

With HDRC, there is no need for aperture and/or exposure-(integration-) time control.

In other words:

> With HDRC, no pixel and no frame can be over- or under-exposed.

In fact, as an accurate photometric sensor, it surpasses the human eye, which still needs pupil control covering about four f-stops to control the range of input intensities.

In the evolution of its senses, the human system has optimised its functionalities for efficiency and sensitivity among other features. The logarithmic conversion function is a wonderful case in point. In Fig. 1.1, we have plotted the OECF of the eye and the HDRC sensor on a comparable scale. Regarding the HDRC sensor, we see that the OECF maps an input range of 8 decades or 28 Bit to an output range of 10 Bit. So the first result of log conversion is information compression from 28 to 10 Bit:

The log converting HDR pixel provides natural bit-rate compression.

We may be concerned that this means a possible loss of information. As we shall see shortly, the opposite is true. We gain contrast sensitivity in darker, shaded parts of a scene. The so-called contrast sensitivity answers the most critical question:

> What is the minimum change of the input grey value, which causes a just noticeable difference (JND) on the output?

The contrast sensitivity tells us, by how many percent we have to change an input value to obtain this JND on the output.

The human eye can distinguish changes in the input luminance, which are a little smaller than 1%, at sufficient light levels. In this region of "photopic" vision, this contrast sensitivity is independent of the input (grey) level as shown in Fig. 1.6. This is the region of the logarithmic response:

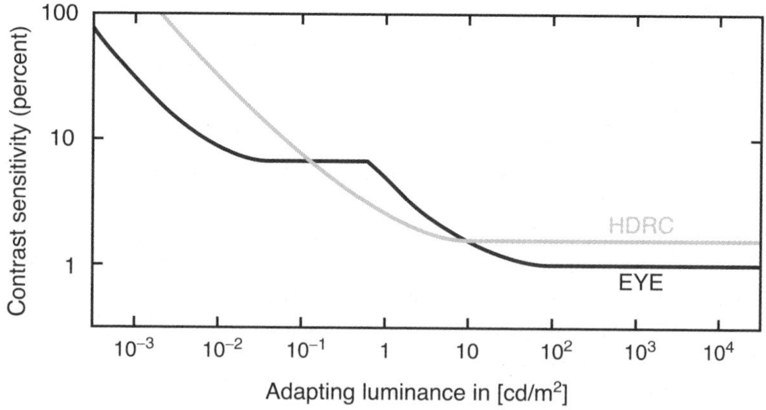

Fig. 1.6. The contrast sensitivity function (CSF) of the human eye and of an HDRC sensor with an input range of 7 orders of magnitude and a 10-Bit output

A logarithmic OECF offers constant contrast sensitivity independent of the input grey value.

It was Weber's finding in the 19th century that a natural response function has this unique feature that a constant "relative" change of the stimulus causes the just noticeable difference in the response. At lower light levels, the contrast sensitivity (CS) of the eye decreases: A relatively larger input change is needed to cause a JND on the output so that the spontaneous sensitivity of the eye is limited. However, with long-time adaptation, the sensitivity range then extends to very low light levels, as shown by the low-light tail of the CSF. We also find in Fig. 1.6 the CSF of the HDRC sensor. In the case of a 10-Bit output covering 7 orders of magnitude on the input, it is 1.5% and constant over 5 orders of magnitude. For the sensor with digital output, the JND is just 1 least-significant bit (LSB), often also called 1 digital number (DN) in the text. The sensitivity decreases at low light levels similar to the eye. However, the sensitivity is better than the spontaneous sensitivity of the eye. In the present discussion, we call sensitivity, in the language of film and standards, the minimum detectable signal. The relatively high performance of the electronic sensors is due to the properties and extensive technology development of silicon (Si) sensors. We put this in perspective relative to ASA and DIN sensitivity standards. For this purpose, we use the relation that a "green" light energy of 8 mlx s produces a just noticeable exposure (or a signal-to-noise ratio of unity) on a 100 ASA film.

The electronic receptor of choice, namely the Si photodiode, is limited in its detection capability by the so-called dark current. Its dominant component in miniature photodiodes is that along the perimeter of the photodiode. To get a feel for its significance, we assume that 50 dark electrons per micron are generated in 25 ms (a frame rate of 40 frames/s^{-1}) at a temperature of 50°C. The statistical variance of this number of electrons, the so-called shot noise, is given by the square root of this number (see Chap. 3). With a green illuminance of 20 mlx we would collect typically .1 electron/μm^{-2} in this frame time. The minimum detectable luminance is reached, when the number of photoelectrons has reached the number of electrons due to the dark shot noise. This would be the absolute physical limit. Other noise sources in an actual Si sensor make the total noise bigger and the sensitivity worse than the ideal presented here in Table 1.2.

This exercise shows that Si sensor technology has the potential for very good ASA/DIN ratings and that there are serious trade-offs between photodiode/pixel size and obtainable sensitivity. Progress in HDR capture in the future depends mostly on progress in low-light sensitivity. This exercise gives us a glimpse at where to look for improvements.

The extended range of high contrast sensitivity down to low light levels is essential for capturing details in the shady parts of HDR scenes. An example is shown in Fig. 1.7.

Table 1.2. The sensitivity of miniature Si photodiodes for different square sizes in terms of ASA and DIN ratings (see text for model assumptions)

Minimum detectable (mlx s): Unity signal-to-noise ratio (shot-noise limit)							
ASA	6,400	3,200	1,600	800	400	200	100
DIN	39	36	33	30	27	24	21
mlx s	0.12	0.24	0.5	1.0	2.0	4.0	8.0
Signal e μm^{-2}	0.25	0.5	1.0	2.0	4.0	8.0	16.0
Photodiode size (μm)	15.0	9.3	5.9	3.7	2.3	1.8	0.9

Reference:
Quantum efficiency: 50%
Temperature: 50°C
Dark electrons: 50e μm^{-1}
Int. time: 25 ms

Fig. 1.7. Portrait illuminated by intense side lighting, captured by an HDRC camera 512 × 256 pixels, 1/60 s, 1996

Inspite of the intense side lighting, the shaded half of the face shows the same amount of detail as the illuminated half, allowing safe contour extraction and face recognition.

We use this image to introduce another powerful feature eye-like HDR log capture:

Separation (or Disregard) of the Illuminant

We perceive real-world scenes predominantly by content (objects, contours, texture, color, motion) and less by actual and changing illumination. This has led to the feature of "Disregarding the Illuminant" in the context of describing the (HVS). In this introduction, we show practical examples where we correct or remove the incidental illumination in images recorded with HDR log-response sensors in real time to obtain image representations either pleasing for our eyes or suitable for robust machine vision.

Figure 1.8 shows four images of the Macbeth Chart recorded with an HDRC camera where the lens aperture was varied over 5 f-stops. Raw images on the left. On the right side, all pixels on the seemingly "over" or "under"-exposed frames have received an offset corresponding to the log of the numbers

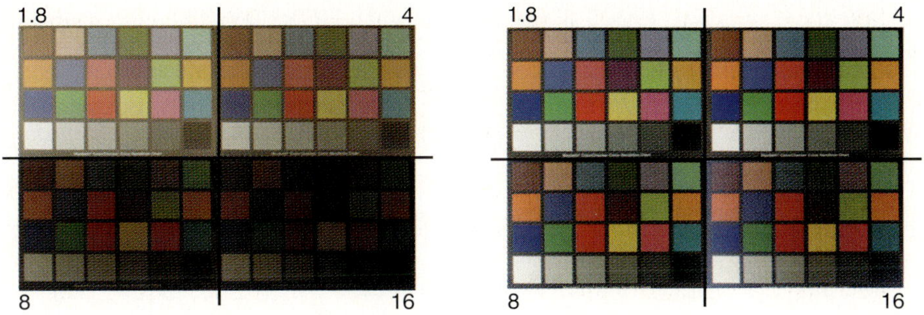

Fig. 1.8. Macbeth chart recorded with an HDRC camera with four apertures. Left: sensor output. Right: Pixel data shifted by common offsets

of f-stops necessary to get what we perceive as the "standard" representation of the chart.

A global "offset" correction, easily done in real time, restores the raw images on the left to the desired images on the right without any further processing or correction.

Leaving details to later chapters, let us consider here that the intensity incident on each pixel of our sensor is the product of the illumination and of the reflectance of that spot in our scene, which is imaged onto this pixel. Because of the logarithmic response of the HDRC pixel, the pixel output is the sum of log-luminance and log-reflectance. Therefore, an offset correction by log-luminance has the desired effect.

The result of this straightforward procedure is shown in Fig. 1.9 for the case of a rather spotty illumination of an object. When the profile of that illumination at the distance of that object, e.g. in front of an ATM, is known for that camera operation, the application of that profile as an offset produces an output image, which comes close to perfect studio lighting and which is ideal for feature extraction and recognition.

Fig. 1.9. One frame from a video sequence of a moving model with spot illumination recorded with an HDRC camera. On the left: The raw frame. In the middle: The log profile of the illumination recorded and stored after the installation of the camera. On the right: The illumination profile added as an offset in real time to the raw frame. Frame rate: 30 frames s^{-1}

We have shown in Figs. 1.8 and 1.9 that with HDRC eye-like video capture we can

Manipulate, Correct or Eliminate the Illuminant.

We just notice here that with the inverse process we can also custom-illuminate HDR log images.

What has been subtly introduced with these HDR color figures is another important feature of eye-like HDR acquisition:

Color Constancy

This phenomenon is the very complex capability of the HVS to perceive a color unchanged whether intensely or weakly illuminated and even if the spectrum of the illuminant changes. The latter is beyond the capability of present HDR capture. However, the first, namely rendering a color of any local area in a scene constant irrespective of whether it is brightly lit or shaded is evidenced by our two figures.

No local or global color correction of HDR log images is needed to compensate the illumination.

The offset by log luminance is the only operation performed to obtain the improved images in Figs. 1.8 and 1.9 Chapter 4 addresses HDR eye-like contrast and color in more detail.

Our paradigm of:

"HDR Photometric Video Capture mimicking the Eye's Response" offers unique features summarised here with references to the further treatment and important applications in this book:

1. Each pixel is a photometric log-compressing sensor with a dynamic range of 8 orders of magnitude (Chap. 2).
2. All pixels in the sensor have the same OECF (Chap. 2) due to calibration in real time.
3. Because of the very high dynamic range, there is no aperture or shutter-(integration) Time control (Chaps. 2, 3, 5, 6, 9 and 10)
4. The sensors have good low-light sensitivity, because the pixels operate basically at the shot-noise limit of their photodiodes (Chap. 3).
5. Constant contrast sensitivity over many decades of grey facilitates robust object detection in high-speed machine vision (Chaps. 5 and 6), in traffic (Chap. 8), in endoscopy (Chap. 9).
6. Pixel output is the sum of log-luminance and log-reflectance (R, G, B) (Chap. 4).
7. Log-luminance is available directly for each pixel (Chaps. 4, 9, 11 and 12).
8. Color per pixel is constant irrespective of luminance level (Chaps. 4, 7, 8, 9, 11 and 12).

The disruptive technology of HDR photometric video capture summarised in this chapter impacts the design of cameras and optics, and it enables entirely new, high-performance vision systems. HDRC video cameras are presented in Chaps. 5, 7, 8 and 9. Lenses for HDR video are discussed in Chap. 6. Some of the most demanding applications of HDR vision were selected for this book.The video-based guidance of aircraft is presented in Chaps. 5 and 6. Chapter 7 is an in-depth treatment of high-speed HDR machine vision for the inspection of complex, highly reflective parts. Chapter 8 offers a perspective on automotive driver assistance and the importance of HDR vision in this context. All the unique features of our new paradigm come to bear in new miniature cameras for minimally invasive endoscopy. It is a very special phenomenon that we return to the impaired eye what we have learned from the healthy eye, namely a subretinal implant, in which HDR log-compressing pixels generate a compatible stimulus for the retinal network as a partial replacement for deceased photoreceptors (Chap. 10).

Direct HDR photometric, log-compressing video capture with its film-like response, log luminance, natural bit-rate compression, constant contrast sensitivity and constant color has formidable implications on the video chain (Fig. 1.10) consisting of coding, transmission, tone mapping and display.

Chapters 11 (tone mapping) and 12 (coding) are critical reviews of these dynamic areas with an emphasis on comparative evaluations and original contributions, which have special affinity to HDR photometric capture. The HDR display presented in Chap. 14 is not only the one with widest dynamic range, but it is also the most illustrative inverse of the separation of log luminance and color in HDR video capture: What is displayed to us is basically the multiplication of regional luminance and high-resolution color.

To conclude this introductory chapter and in view of the many disciplines coming together in this book, we point the reader to the glossary of terms and abbreviations as well as to the comprehensive index in the back.

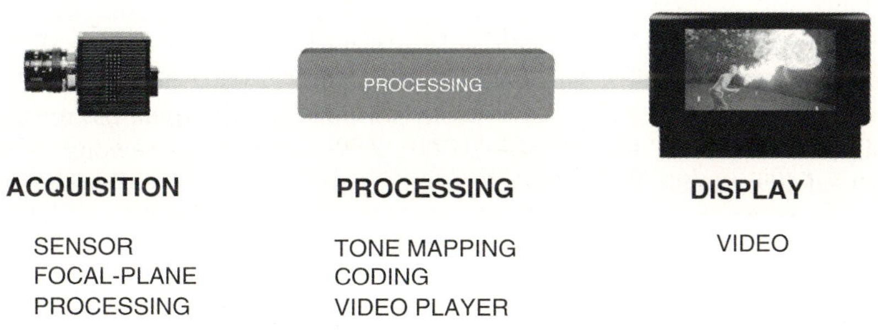

Fig. 1.10. The HDR vision chain from video capture through processing to display

In the case of a logarithmic OECF

$$y = a \cdot \ln x, \tag{2.1}$$

$$\Delta C = \frac{\Delta y}{a} = 1/a, \tag{2.2}$$

if $\Delta y = 1$ LSB in the case of a digital output.

If we have a digital output of N bit, we can trade ΔC against the dynamic range of M decades of input intensity:

$$\Delta C = \frac{1}{a} = \frac{M \cdot \ln 10}{2^N - 1} \tag{2.3}$$

From Table 2.1, we learn the power of the logarithmic OECF:
Input dynamic range: M Decades
Output Signal Range: N Bit

Contrast sensitivity is the percent input change, which will be noticed as a change of 1 LSB (1 DN) in the output.

A dynamic range of $M = 6$ orders of magnitude or 1 Mio.:1 corresponding to 20 bit of optical input is mapped to an electronic output with a range of $N = 12$ bit resolving 0.35% changes of the optical input over the full input range from white to dark. For a 10-bit output, a change of 1.4% would produce 1 LSB. This example of a logarithmic OECF is therefore also said to offer a compression of a 20 bit input to a 10 bit output.

In any real-world system, as the incident optical intensity x decreases, eventually our electronic output signal y will settle to $y \to 0$ for $x \to 0$ as shown in Fig. 2.3.

We can write this OECF as

$$y = a \cdot \ln(1 + x). \tag{2.4}$$

The extrapolation of the log characteristic with the slope a for larger inputs $x \gg 1$ intersects the x-axis at $x = 1$.

Figure 2.3b shows the contrast-sensitivity function (CSF) ΔC for this OECF.

Table 2.1. Contrast sensitivity (percent) of a logarithmic OECF

Output N Bit Input M decades	8 bit	9 bit	10 bit	11 bit	12 bit
2 Dec	1.8	0.90	0.45	0.225	0.112
3 Dec	2.7	1.35	0.675	0.336	0.168
4 Dec	3.6	1.8	0.900	0.45	0.225
5 Dec	4.5	2.25	1.175	0.56	0.280
6 Dec	5.4	2.7	1.350	0.67	0.336
7 Dec	6.3	3.15	1.575	0.784	0.392
8 Dec	7.2	3.6	1.80	0.9	0.45

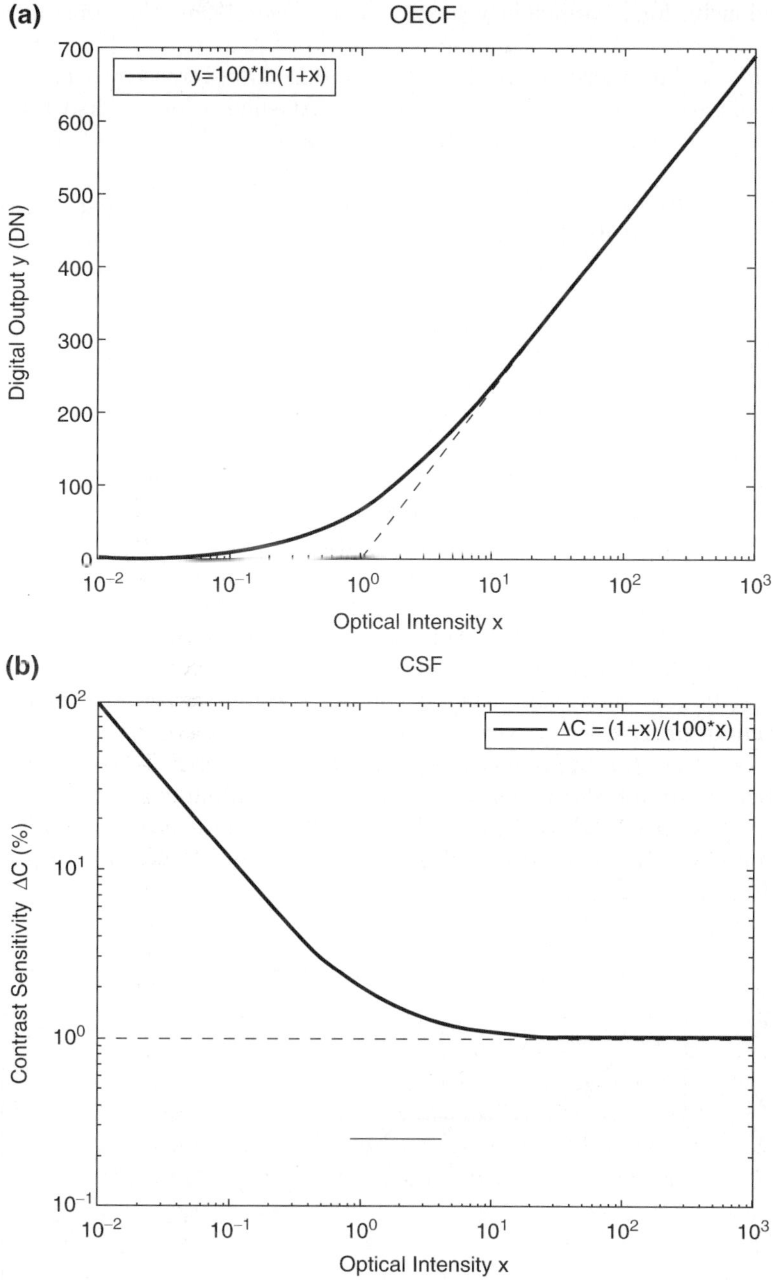

Fig. 2.3. (**a**) OECF of a natural eye-like image sensor. (**b**) CSF of a natural eye-like image sensor

Obviously, for x sufficiently larger than 1, we obtain the constant value $1/a$.

At $x = 1$, ΔC increases to $2/a$. At $x \approx 0.1$, ΔC increases to $11/a$, rising to 1 in our example at $x = 1/a$ for $a \gg 1$. At such a low optical input, we would have to double the input intensity for a noticeable change of 1 LSB in the output.

For log-response sensors we will often plot the OECF as $y = f(\log x)$ (Digits = DN) and its derivative

$$S = \frac{\mathrm{d}y}{\mathrm{d}\log x}\,(\mathrm{DN\,Decade}^{-1}),$$

$$S = \frac{x \cdot \mathrm{d}y}{\log e \cdot \mathrm{d}x}.$$

To convert S into the contrast sensitivity

$$\Delta C = \frac{100}{x \cdot \mathrm{d}y/\mathrm{d}x}(\%),$$

we observe that $S = (\Delta C \cdot \log \mathrm{e})^{-1}$. Therefore, we represent the CSF either as $S\,(\mathrm{DN\,Decade}^{-1})$ or $\Delta C\,[\%]$. For example, a contrast sensitivity of 1 or 2% is equivalent to 230 or 115 DN Decade^{-1}, respectively.

In Fig. 2.4, we show, somewhat simplified, the contrast-sensitivity function of the human eye.

We see that for luminances higher than $100\,\mathrm{cd\,m}^{-2}(3.4\,\mathrm{lx})$ its contrast sensitivity is about 1% and independent of the luminance. The spontaneous sensitivity is only 10% at $1\,\mathrm{cd\,m}^{-2}(34\,\mathrm{mlx})$. The characteristic point, where it has doubled to 2% occurs at about $10\,\mathrm{cd\,m}^{-2}$, which we will call the 3 db point from now on.

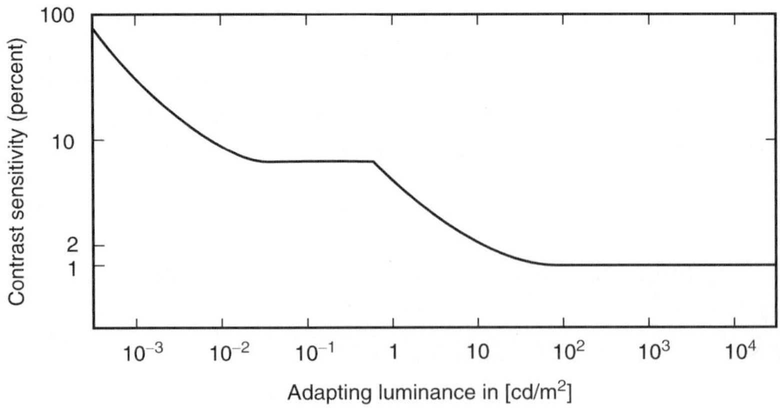

Fig. 2.4. The CSF of the human eye

What Fig. 2.4 tells us is that in human vision minimum perceivable differences are about 1% over several decades of luminance at levels higher than 10 cd m^{-2} and, that in this range, which is also called the photopic response, the human eye has a logarithmic OECF. Its spontaneous response is limited at about 0.3 cd m^{-2} (10 mlx), reaching out to mcd m^{-2} with long-term adaptation.

We will use the eye characteristic of Fig. 2.4 as nature's guidance to the electronic implementation of high-dynamic-range (HDR) image sensors.

2.2 The HDRC (High-Dynamic-Range CMOS) Pixel

Silicon integrated-circuit technology has been the basis of electronic image sensors for over 30 years for two dominant reasons:

1. Silicon photodiodes are the best visible-light detectors.
2. Metal Oxide Silicon (MOS) technology facilitates the storage and loss-free transport of photo-generated electronic charges.

The pn junction of silicon photodiodes can act efficiently as converter of visible-light photons into electronic charges. The quantum-efficiency and spectral-sensitivity diagram (Fig. 2.5) shows that for each photon absorbed in the pn junction almost one electron can be obtained. Furthermore, silicon integrated-circuit technology as the world's dominant microelectronic manufacturing technology allows the closely spaced integration of millions of such photodiodes on a single chip. The separation and isolation of these diodes in array-type image sensors has been and is continuously optimized to reduce leakage currents, which are the primary cause for the detection limit. Leakage-current figures of merit as low as 10 pA cm^{-2} have been realized at room temperature. The currents have a characteristic dependence on temperature where silicon provides the best compromise among all semiconductors.

A rule of thumb for this temperature dependence is that leakage currents double every 8–10°C in the temperature range –40 to +100°C.

The generated photo charges emerge from the photodiode as a photocurrent, which is strictly proportional to the incident photon flux over many orders of magnitude of incident intensity.

In spite of near-ideal optoelectronic conversion, the generated photocurrents are exceedingly small.

Table 2.2 provides a listing of some useful relations and quantities. Based on these numbers, a photodiode with an area of 25 μm^2 illuminated by moonlight of 10 mlx will offer a photocurrent of 0.13 fA. This is equivalent to just 800 e s^{-1}. Such small currents cannot be sensed directly as currents, and direct current amplifiers at these levels are also not feasible in mainstream microelectronics. This mainstream is MOS technology. For reasons of power consumption and noise immunity, complementary MOS (CMOS) technology

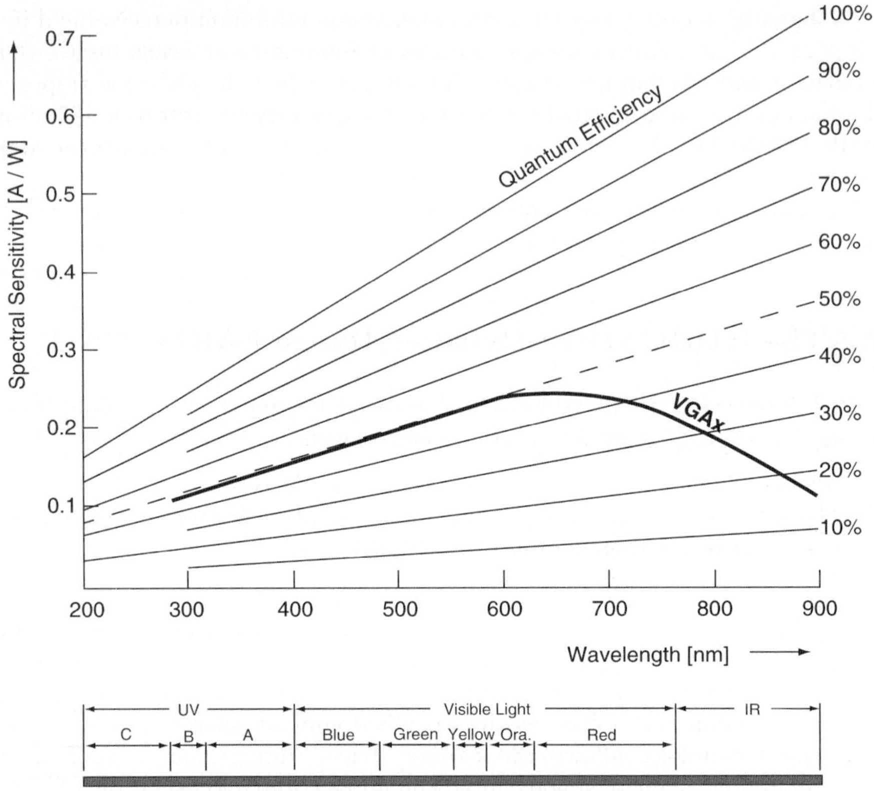

Fig. 2.5. Quantum efficiency and spectral sensitivity vs. the wavelength of light. Also shown: The characteristic of an integrated silicon photodiode with an area of 40 μm^2

Table 2.2. Parameters of a reference pixel

Pixel area	(μm^2)	7.4 × 7.4
Photodiode (PD) area	(μm^2)	5.0 × 5.0
Capacitance	(fF)	2.5
Dark charge	(e)	20
Dark shot noise	(e)	4.5
Conversion efficiency	(e mlx^{-1})	2
	(DN mlx^{-1})	0.33
Sensitivity	(V lx^{-1} s^{-1})	5.4
Reset noise	(e)	20
Quantization (1 DN)[a]	(e)	6
	(mV)	0.4
Conditions: $T = 20°$C, $T_{\text{int}} = 25$ ms		

[a]Linear 10 bit

has a share of over 90% of the integrated-circuit world market today. In MOS integrated circuits, we manipulate tiny electronic charges on tiny capacitors, measure the voltage changes on those capacitors and subsequently amplify and digitize these voltages. The key active component facilitating these operations is the MOS transistor. It functions not only as an ideal voltage-controlled current amplifier, it is also a voltage-controlled switch with a very high-off resistance and a very large on–off resistance ratio. The control gate of the MOS transistor can also serve as a storage capacitor. As soon as millions of such transistors per chip were realized as mass products in the eighties of the twentieth century their combination with photodiodes became a major development area in what has since been called active-pixel image sensor (APS).

For HDR sensing, we focus on three observations:

1. We stated before that the signal type of modern microelectronics is a voltage signal.
2. The optical input x, namely photon flux, is converted in the silicon photodiode into a photocurrent with strict proportionality
3. The electronic output signal y of our pixel should be a voltage and the optoelectronic conversion function (OECF) should be logarithmic and, therefore, a pixel should convert the photocurrent x into a voltage y, which would be proportional to the logarithm of the photocurrent x:

$$y = a \cdot \ln(1 + x). \tag{2.4}$$

How can we accomplish this?

Figure 2.6 shows an NMOS transistor, basically a four-terminal device. The current of interest is the drain current ID. We consider two voltages, the gate-to-source voltage VGS and the drain-to-source voltage VDS. Dependencies between these voltages and the drain current in the predominant modes of operation of the MOS transistor have either linear or parabolic relationships. However, for small control voltages VGS the drain current decreases or increases exponentially as VGS is reduced or increased. This is normally considered an unfavorable "leakage" or "parasitic" current. However, a very well-established and controlled physical mechanism is responsible for this relationship.

Electrons from the source have to cross a barrier towards the channel and the collecting drain electrode. This barrier height is modulated by VGS and VDS according to the following relationship [1]:

$$\mathrm{ID} \approx I_0 \exp[(A.\mathrm{VGS} + B.\mathrm{VDS})/nV_t], \tag{2.5}$$

where A and B depend on the length of the transistor channel, the thickness of the gate oxide, the doping of the transistor channel and of the transistor well and on VBS. Representative values are $A = 0.6$ and $B = 0.2$. A typical value for n is 1.5 and V_t is the "thermal" voltage kT/q, which is 26 mV at

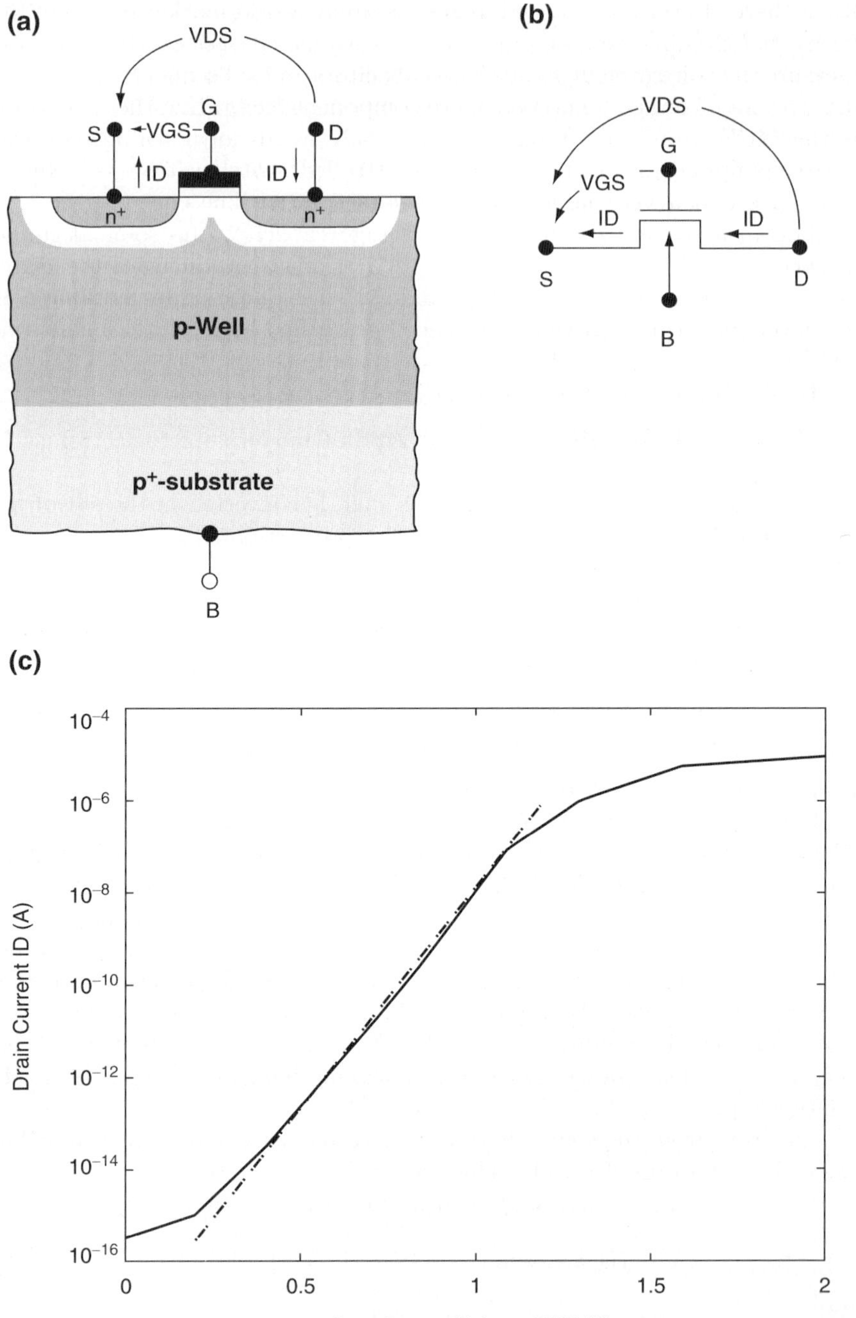

Fig. 2.6. NMOS transistor: (**a**) cross section, (**b**) symbol, (**c**) typical ID (VGS) characteristic for VGS = VDS

$T = 300K$. A typical characteristic is shown in Fig. 2.6 for the conditions VGS = VDS and VBS = 3.0 V.

Evidently, this transistor provides a drain-source voltage VDS, which is the precise logarithm of the drain current ID over at least 8 orders of magnitude. In this case, the transistor actually operates as a highly nonlinear resistor. The essential barrier control (2.5) is provided by the voltage difference between gate and source and the gate is actually connected to the drain: VGS = VDS.

The photodiode current is injected at the source electrode of the MOS transistor T1 in the pixel diagram Fig. 2.7.

In fact, the source area of the transistor can be extended to become the photosensitive cross-section of the pixel. In this fashion, contacts are saved and density is gained in the pixel cell. The output signal voltage is available at the source of T1 and this voltage becomes the gate control voltage of a second transistor T2, which provides the charge and discharge current for the read-out electronics of the sensor.

A select transistor T3 operating as a switch completes the pixel cell as shown in Fig. 2.7. This figure also contains a cross-section through the pixel,

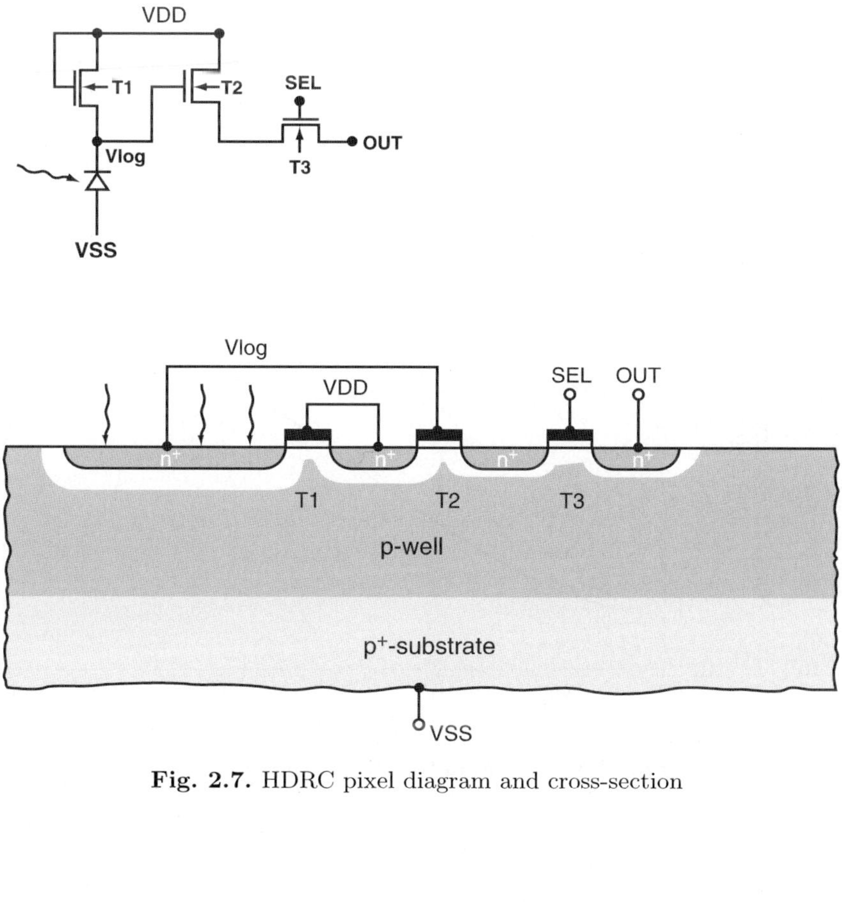

Fig. 2.7. HDRC pixel diagram and cross-section

which shows the P- and N-type regions of the diode and the transistors as well as all the gates, contacts and interconnects needed.

A HDRC sensor with 64 × 64 pixels of this type was first realized and published in February 1992 [2] and cameras were then tested on vehicles to record high-dynamic-range scenes on the road including tunnels.

For the performance assessment of the HDRC pixel, we consider a reference pixel, which has the parameters given in Table 2.2. The simplicity of the HDRC pixel as shown in Fig. 2.8 allows the direct measurement of the current in the branch of the pixel consisting of the series connection of the log transistor T1 and the photodiode as a function of the illumination (Fig. 2.8).

This data is obtained from measurements on a 768 × 496 pixel HDRC sensor under flat-field illumination. On the straight-line section, we find a slope of 15 fA lx^{-1}. This is equivalent to a spectral sensitivity of 0.3 A W^{-1}. The voltage signal V log is obtained basically after a string of unity-gain amplifiers (pixel follower T2, mux1, and output amplifier) at the output of the sensor, and measured results are shown in Fig. 2.9.

This is the OECF of the HDRC sensor.

This data spans more than 7 orders of magnitude in the semi-log diagram. It shows the perfect logarithmic behavior and an overall shape similar to the model OECF as shown in Fig. 2.3. The output is given both as a voltage (mV) and as digital numbers (DN) with a 10 bit range. We can readily extract the CSF as given in Fig. 2.9b, which is the slope of Fig. 2.9a.

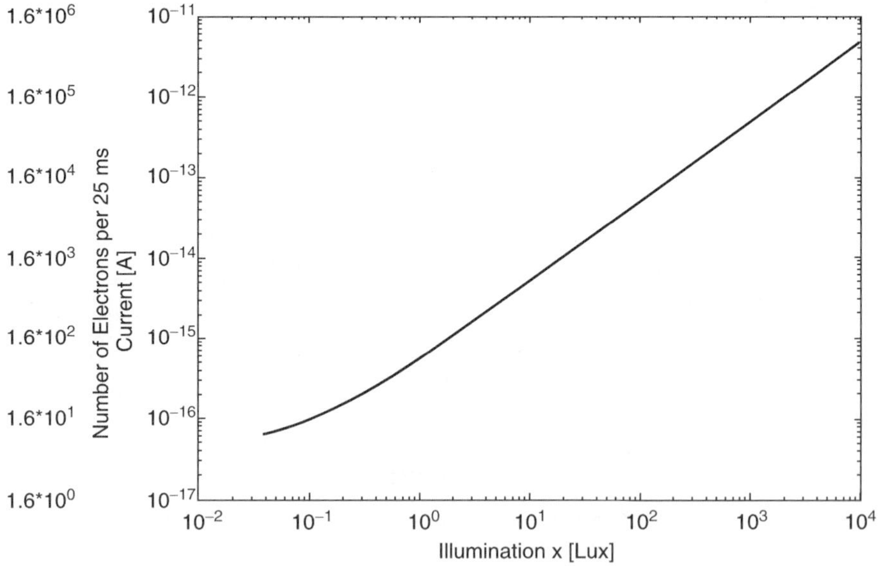

Fig. 2.8. Current in the HDRC 10 μm pixel

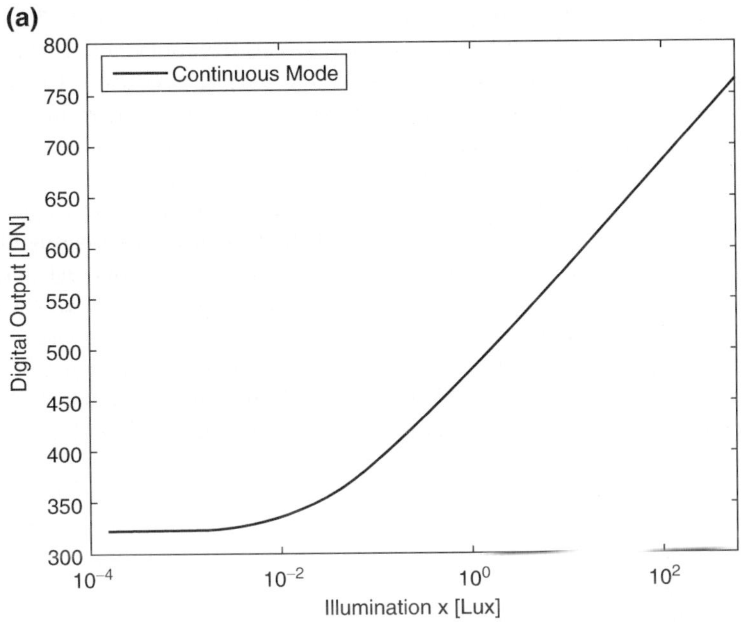

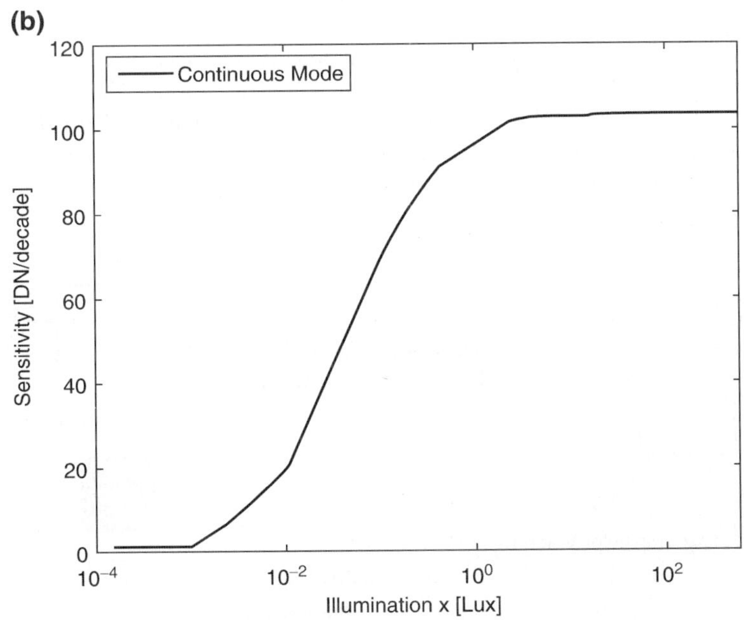

Fig. 2.9. (**a**) The OECF of the HDRC 10 μm pixel (**b**) The CSF of the HDRC reference pixel

In this setting, the output signal range covers more than 7 orders of magnitude or 140 dB, and the contrast sensitivity obtained is 1%, which corresponds to our expectations for a 10 bit output according to Table 2.1. The characteristics show that the HDRC pixel, as simple as it is, has the capability of recording an extremely wide dynamic range. As we see in Fig. 2.9, our reference pixel has a leakage current of 0.5 fA resulting from the photodiode and from the log-node contact. The photocurrent is equal to this "dark" current at the 3 dB point of the OECF, which we find has this magnitude at an illumination of 20 mlx. At this point on our OECF, the contrast sensitivity deteriorates by a factor of 2 compared with the log-response part of the OECF.

At 2 mlx, the contrast sensitivity is 1/10 of the sensitivity in the log-region or, in other words, a grey level change at this low illuminance would have to be 10% to be just noticeable. The OECF in Fig. 2.10 of our reference HDRC pixel has the ideal model shape according to (2.4). The output signal y with a DN range 0–1,023 has a contrast sensitivity of 110 DN decade^{-1} of illumination.

The HDRC pixel realizes the OECF of the human eye and, when we compare the critical 3 dB points, we find that the electronic pixel with a 3 dB point of 0.6 cd m^{-2} (20 mlx) performs better than the human eye where this point is at about 10 cd m^{-2} (340 mlx) for the spontaneous sensitivity.

We should also point out here that the HDRC pixel sensitivity and minimum detectable level are basically independent of any exposure or integration time. The pixel follows the photocurrent and its changes continuously in the basic mode of HDRC operation. We can read the voltage Vlog randomly at any time by accessing the pixel and reading out the voltage VOUT.

Having shown that the HDRC pixel has the ideal logarithmic eye-like conversion function, we move on to study a HDRC array sensor.

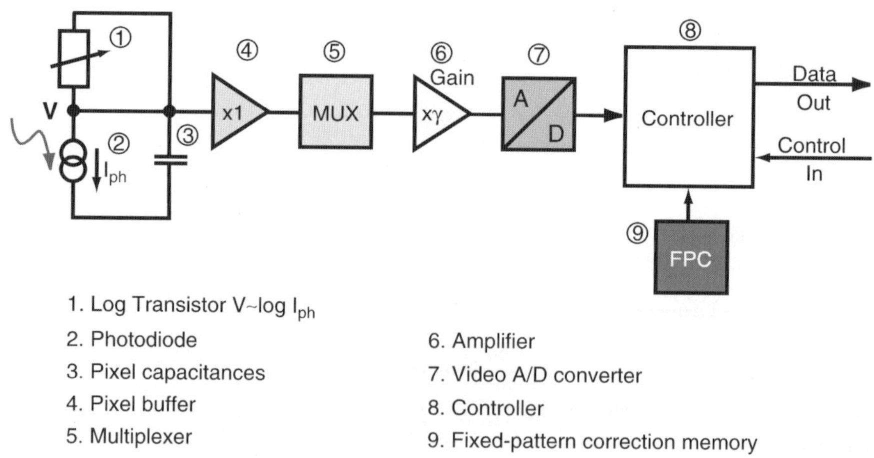

Fig. 2.10. Block diagram of HDRC sensor

2.3 The HDRC Sensor

The basic HDRC pixel as shown in Fig. 2.7 does not need any clock or reset signals. It monitors the optical input (the photocurrent) continuously. The dynamic range is so wide that no camera aperture or shutter time adjustments are necessary. This means that the HDRC pixel basically does not need any control or adjustment loops with critical timing requirements. In fact, the HDRC pixel never "blinks", that is to say, it cannot be caught off guard, it cannot be over- or underexposed, it gives a continuous ready-any-time response to its optical input. This fact also makes the HDRC pixel very easy to integrate in a large-array image sensor with full random access to any pixel any time. The basic block diagram of an HDRC sensor is shown in Fig. 2.10.

This sensor requires just one clock signal and a single voltage supply. It operates in the progressive scan mode with pixels being read out synchronous with the chip clock. Again, there are no control loops and restrictions on read-out. The pixel access time is constant, in the reference sensor it is 80 ns.

The systematic characterization of this HDRC sensor starts with the sensor output under flat-field illumination.

Figure 2.11 shows an overlay of the conversion functions of the 400,000 individual pixels in our reference sensor. The result is a medium characteristic

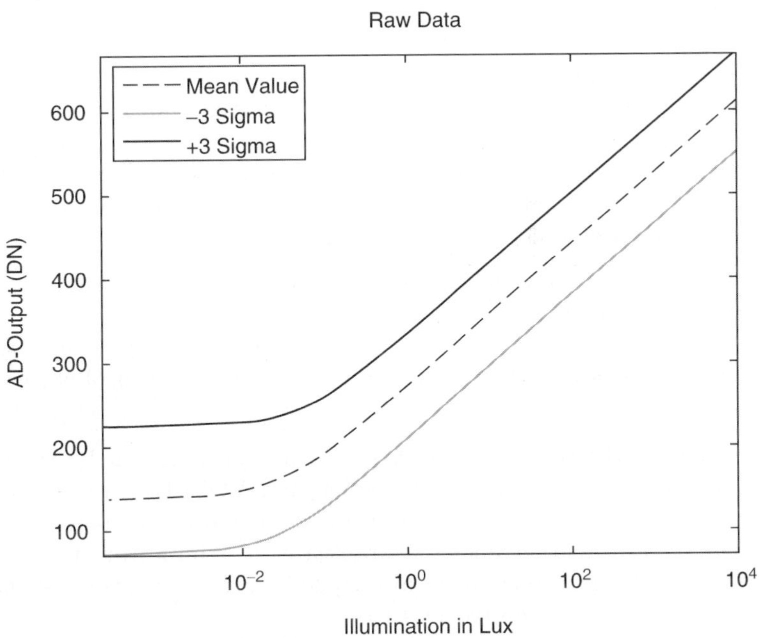

Fig. 2.11. Overlay of the OECFs of the 400,000 pixels of a 768 × 496 pixel HDRC sensor

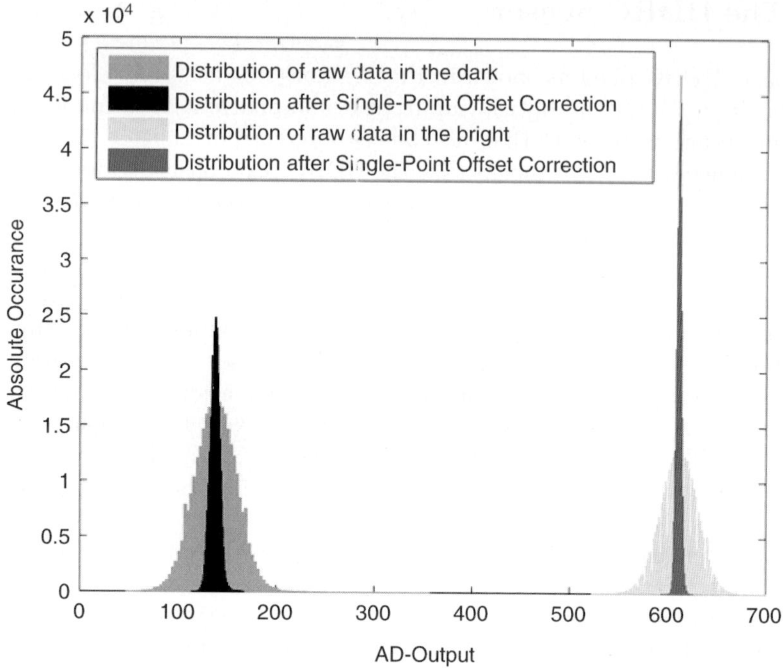

Fig. 2.12. Dark and bright histogram of the sensor fixed pattern

similar to the one shown in Fig. 2.10 with a corridor around it, showing deviations of individual pixels from the mean. The analysis and corrections of this fixed pattern is important in order to obtain high-quality image acquisition.

Figure 2.12 shows a dark histogram and a bright histogram for the fixed pattern. An effective fixed-pattern correction (FPC) requires an understanding of its causes. Two independent phenomena are present:

1. The distribution at low light levels is dominated by the distribution of the dark currents of the pixel photodiodes.
2. The distribution at bright levels is caused by distributions of transistor parameters, mostly the distribution of their so-called threshold voltages VT.

In either region, the deviations from the mean function can be treated as voltage offsets, VOFFD in the dark section and VOFFB in the bright section. Because of the different causes of these distributions, namely, either the photodiode or on the other hand the transistor, it is important to note that these offsets are not correlated within a pixel. In fact, a pixel may have a positive VOFFD and a negative VOFFB:

This means that two or more correction values have to be stored per pixel and the selection between the two should take place at a threshold near to the

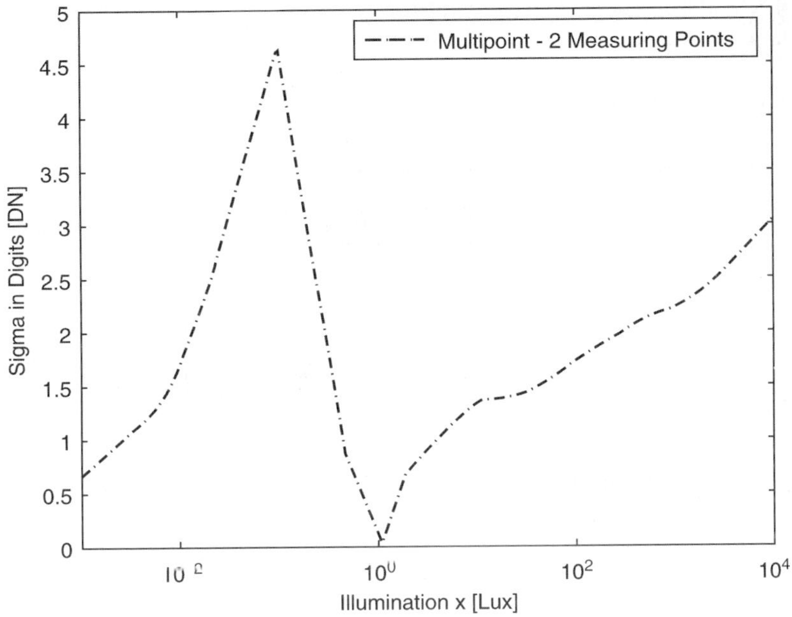

Fig. 2.13. Standard deviation after fixed-pattern correction with VOFFD and VOFFB per pixel

3 dB-point of the conversion function. If such a correction is implemented, the overlay of the response curves is shown in Fig. 2.13.

At this correction level, the standard deviation from the mean OECF is reduced to better than 3 DN in the range from 1 mlx to 5 klx. More refined fixed-pattern correction will be described in the following section:

For the HDRC sensor, adding an offset per pixel enables several other powerful functions besides fixed-pattern correction, and it teaches us some unique properties of logarithmic sensor outputs.

In Fig. 2.14, we show that we introduce the offset memory as a digital memory behind the analog-to-digtal converter. Digital storage and arithmetic are robust and exceedingly cost-effective. We also show the camera lens and the scene illumination as important components of our imaging system. Local optical intensity $O(\lambda, x, y)$ incident on pixel (x, y) is the product of many components in the path of light rays falling on this pixel:

The illuminance $L(\lambda, x, y)$ illuminating that part of the imaged object
The reflectance Refl (λ, x, y) of that part of our imaged object
The local transmittance $T(\lambda, x, y)$ of the lens and coatings effective for the pixel (x, y)
The spectral sensitivity $\alpha(\lambda, x, y)$ of the photodiode (x, y).

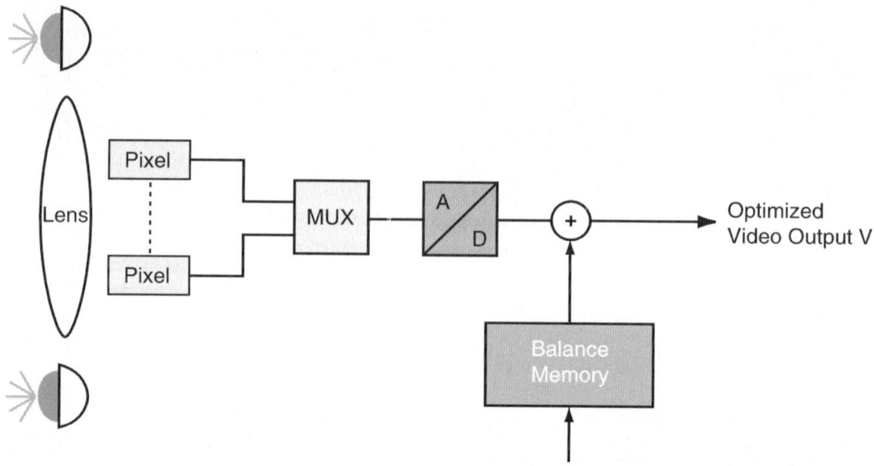

Fig. 2.14. Block diagram with digital offset correction memory

As a result, the optical intensity falling on our pixel is

$$O(\lambda, x, y) = L(\lambda, x, y) \times \text{Refl}(\lambda, x, y) \times T(\lambda, x, y). \tag{2.6}$$

This intensity produces a photocurrent in the pixel

$$I_{\text{ph}}(x, y) = \alpha(\lambda, x, y) \times O(\lambda, x, y). \tag{2.7}$$

If the signal is big enough so that we are in the logarithmic section of our conversion function, the sensor output voltage $V(x, y)$ for pixel (x, y) will be

$$\begin{aligned} V(x,\, y) &= a \log I_{\text{ph}}(x,\, y) \\ &= a(\log L + \log \text{Refl} + \log T + \log \alpha) + FP(x,\, y), \end{aligned} \tag{2.8}$$

where $\text{FP}(x, y)$ is the fixed-pattern offset of pixel (x, y).

Equation (2.8) can be considered as the master equation of HDR logarithmic imaging. Let us first look at the effect of illuminance and aperture and let us assume that the illuminance is uniform over our observed scene. We see that, in the HDRC-sensor output, illuminance as set by the lens aperture comes in just as an offset.

If the aperture is changed by one f-stop (a factor of 2) or if the illuminance changes by a factor of 2, the HDRC sensor output will change be an offset of $a \log 2$, which is a small change if we have a dynamic range of 27 f-stops.

On the other side, it means that any recording with the wrong aperture can be ajusted by just an offset to obtain the desired output.

Figure 2.15 is an example of these effects. The grey chart has been recorded with varying the camera aperture from f 1.4 to f 16 that is by 7 f-stops. In Fig. 2.16a the recorded grey levels change significantly. However, the contrast

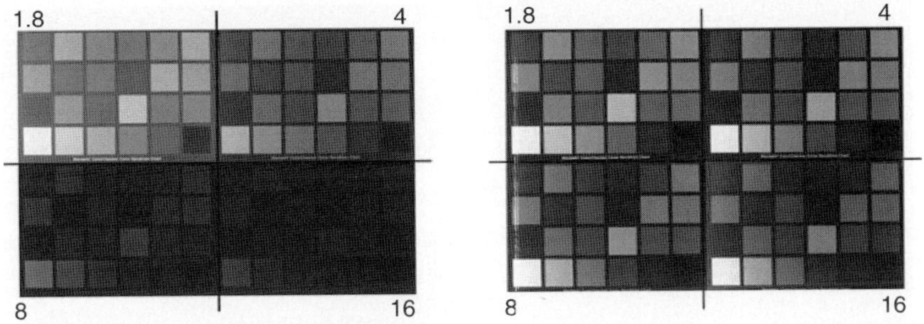

Fig. 2.15. Grey chart recordings with apertures from $f1.4$ to $f16$. Left: Raw digital image. Right: Images after a digital offset proportional to log aperture

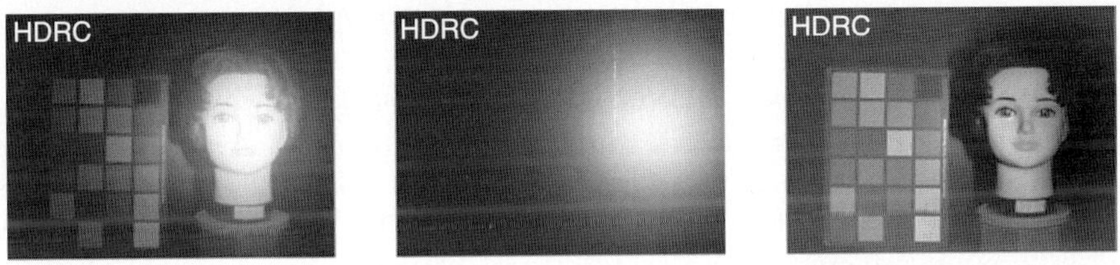

(a) Original HDRC® Frame **(b)** Log Illuminance (Spotlight) **(c)** HDRC® Frame after Log Illuminance Offset

Fig. 2.16. Scene with spotlight illumination: (**a**) raw HDRC from output. (**b**) HDRC data equivalent to the log of the illumant (**c**) "Standard Appearance" after subtracting the illuminant

steps are well resolved even at $f : 16$. In Fig. 2.16b, the four images are shifted by the appropriate offsets resulting in identical grey-chart appearance.

Now, consider a scene illuminated by a spotlight, that is with a highly non-uniform illumination. This is quite apparent in the face imaged in Fig. 2.16a.

However, we recorded the illuminant distribution in the plane of the face with putting a white surface there and we obtained the distribution of Fig. 2.16b. If we subtract Fig. 2.16b from Fig. 2.16a, we obtain the image in Fig. 2.16c, which now has the appearance we would get from an elaborate balanced studio illumination. This illuminance equalization is quite straightforward and very effective in rendering objects with their characteristic appearance not distorted by changing lighting conditions. The recording also shows that, in spite of the spotty illumination, bright regions are free from white saturation, that rich detail is available in deep shades and that we can pull up this information with illuminance equalization performed as a simple offset operation.

We also notice that wavelength – dependent sensitivity α (Red, Green, Blue) of the pixel can be easily adjusted by an offset contrary to a pixel with a linear aperture function, where we would have to apply multiplicative corrections.

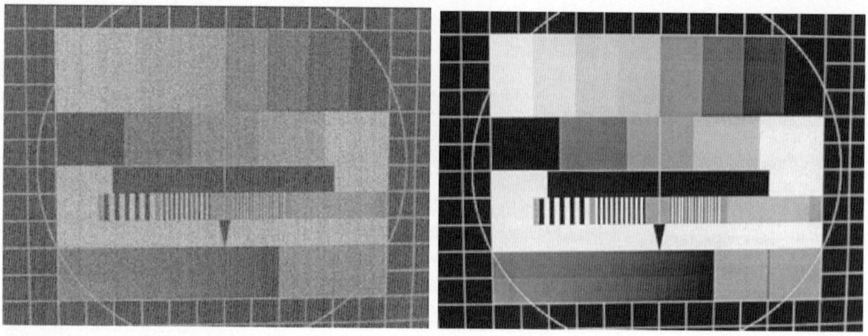

Fig. 2.17. Left: Scene taken with the HDRC sensor without any fixed-pattern correction. Right: Scene taken with the HDRC sensor with fixed-pattern correction

The exposure of the HDRC sensor to optical input with a spatially high dynamic range ratio raises concern about crosstalk between neighboring pixels. Measurements have shown the second and third neighbor of a brightly illuminated pixel record to be just 1% and 0.03% of that intensity. These results are satisfactory considering other factors limiting the spatial resolution of image features.

2.4 Fixed-Pattern Correction of HDR Imagers

Due to the production process, each pixel has a different OECF with different parameters. This results in a speckled image when illuminating a CMOS image sensor with homogeneous light. Recording normal scenes with the HDRC sensor, fixed-pattern noise degrades the image to an extend that the scene can hardly be recognized (see Fig. 2.17). Without corrective measures the image will have low contrast. Consequently, in order to use HDR imagers, fixed-pattern noise must be corrected.

2.4.1 Physical Background of Logarithmic OECF

The OECF used for the fixed-pattern correction algorithm is more complex than the one previously mentioned. In order to achieve high-quality correction results, it is important to consider the following representation of the OECF:

$$y = a + b \cdot \ln(x + c). \qquad (2.9)$$

Parameter a presents the offset, b the gain and c the illumination at the 3 dB-point. In order to understand the present approach, it is important to know the physical behavior of the pixel. To get a logarithmic OECF, the conversion transistor must work in weak inversion, defined by the following relationship [3]:

$$I_{\mathrm{D}} \approx I_{\max} \cdot \frac{V_{\mathrm{GS}} - V_{\mathrm{thresh}}}{n \cdot V_t}. \quad (2.10)$$

I_{D} represents the drain current, $I_{\max}$ the maximum current in weak inversion, V_{GS} the gate-to-source voltage, V_{thresh} the threshold voltage, n the subthreshold slope factor and V_t the thermal voltage (26 mV at 300 K). The threshold voltage defines at which point the transistor changes from weak inversion into strong inversion mode.

The pixel produces a photocurrent out of the incident photons, which is then transformed into a voltage. The total current consists of the photocurrent and the dark current.

$$I_{\mathrm{D}} = I_{\mathrm{Dark}} + I_{\mathrm{Photo}}. \quad (2.11)$$

As the total current corresponds to the drain current and the gate–source–voltage to the output, (2.10) has to be changed as follows:

$$V_{\mathrm{GS}} = n \cdot V_{\mathrm{T}} \cdot \ln\left(\frac{I_{\mathrm{Photo}} + I_{\mathrm{Dark}}}{I_{\max}}\right) + V_{\mathrm{thresh}}. \quad (2.12)$$

Equation (2.12) can be further transformed

$$\begin{aligned} V_{\mathrm{GS}} &= a_{\mathrm{ph}} + b_{\mathrm{ph}} \cdot \ln(I_{\mathrm{Photo}} + c_{\mathrm{ph}}), \quad (2.13)\\ a_{\mathrm{ph}} &= V_{\mathrm{thresh}} - \ln(I_{\max}) \cdot n \cdot V_{\mathrm{T}},\\ b_{\mathrm{ph}} &= n \cdot V_{\mathrm{T}}\\ c_{\mathrm{ph}} &= I_{\mathrm{Dark}}. \end{aligned}$$

As (2.13) is based on currents and voltages, new parameters a_{ph}, b_{ph} and c_{ph} are defined. The extended OECF of (2.9) corresponds to them. This means that offset a of the OECF mainly depends on the threshold voltage and gain b on the subthreshold slope. The illumination at the 3 dB-point is directly connected to the dark current.

2.4.2 Parameter Extraction with Software

In order to get parameters a, b and c the sensor is measured with three different illuminations, the first one in the dark, e.g. 0.001 lx. This point is especially important to calculate parameter c. The medium illumination at, e.g., 10 lx and the bright one at, e.g., 10,000 lx are needed to define parameters a and b. Following the measurements, the data for each pixel is used to get the individual parameter set by using the following equations:

$$\begin{aligned} b &= \frac{y_{\mathrm{bright}} - y_{\mathrm{medium}}}{ln(x_{\mathrm{bright}}) - ln(x_{\mathrm{medium}})}, \quad (2.14)\\ a &= y_{\mathrm{bright}} - b \cdot ln(x_{\mathrm{medium}}),\\ c &= \mathrm{e}^{\frac{y_{\mathrm{dark}} - a}{b}} - x_{\mathrm{dark}} \approx \mathrm{e}^{\frac{y_{\mathrm{dark}} - a}{b}}. \end{aligned}$$

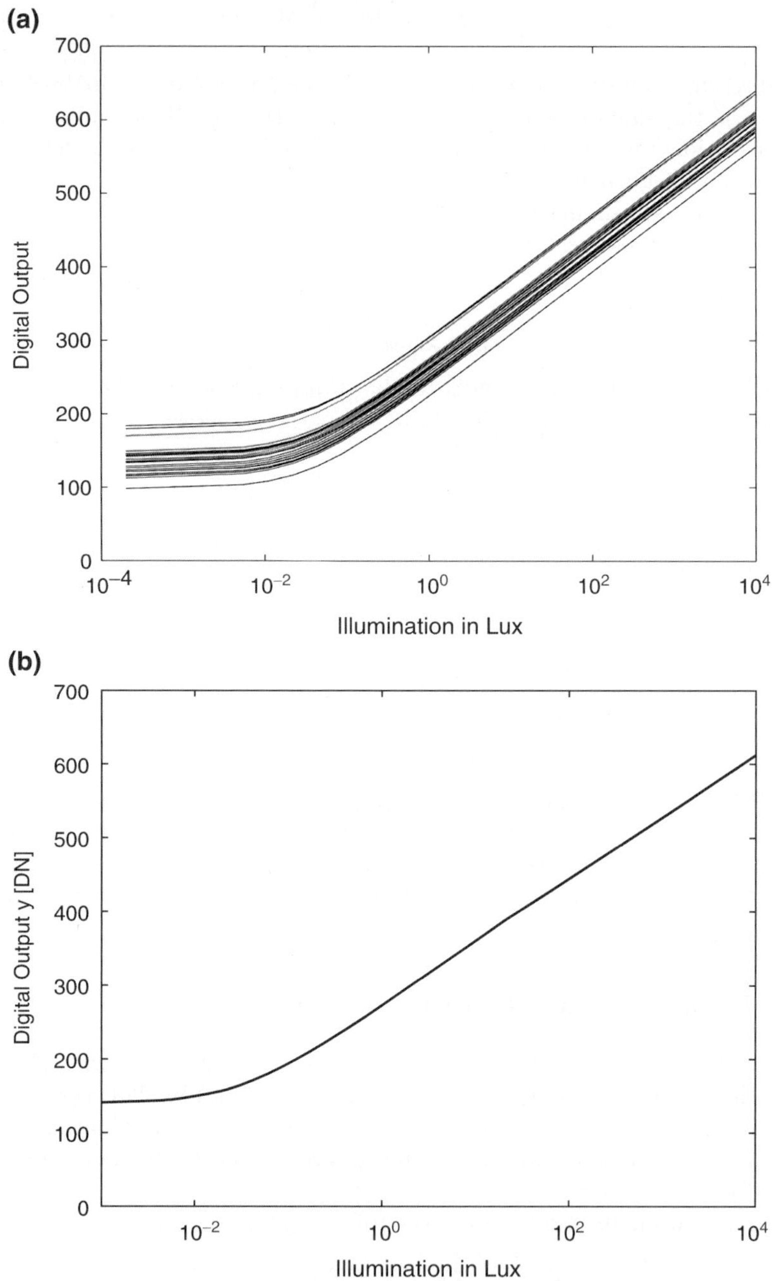

Fig. 2.18. (**a**) Individual pixel characteristic without correction shown for 35 pixels (**b**) Individual pixel characteristic with correction shown for 35 pixels

In these expressions x always represents the illumination, e.g., $x_{\mathrm{dark}} = 0.001\,\mathrm{lx}$ and y_{dark} is the corresponding output value of the pixel.

Figure 2.18a shows that without any correction all measured 35 pixel cells have a different characteristic curve. The aim of the correction is to get, in the optimum case, one single output characteristic shown in Fig. 2.18b. In order to get such a reference curve, the algorithm takes the mean values of the individual parameters a_{i}, b_{i} and c_{i}. Finally, the correction is done with the inversion function (2.15).

$$f^{-1}(y) = \bar{a} + \bar{b} \cdot \ln\left(\mathrm{e}^{\frac{y_{roh} - a_i}{b_i}} - c_i\right), \tag{2.15}$$

$$y_{soll} = \bar{a} + \bar{b} \cdot \ln\left(\mathrm{e}^{\frac{y - a_i}{b_i}} - c_i + \bar{c}\right).$$

2.4.3 Effects of Parameter Variation on the OECF

The following figures show the effects of changing these parameters independently.

Figure 2.19 demonstrates the importance the variation in illumination plays in the three parameters: parameter a is always important. This means that the fixed-pattern correction algorithm is always a decisive factor. If parameter b varies between the pixels, fixed-pattern noise is mainly caused in bright conditions (see Fig. 2.20). The opposite effect can be seen when changing the 3 dB-Point. This causes high noise in the dark (see Fig. 2.21).

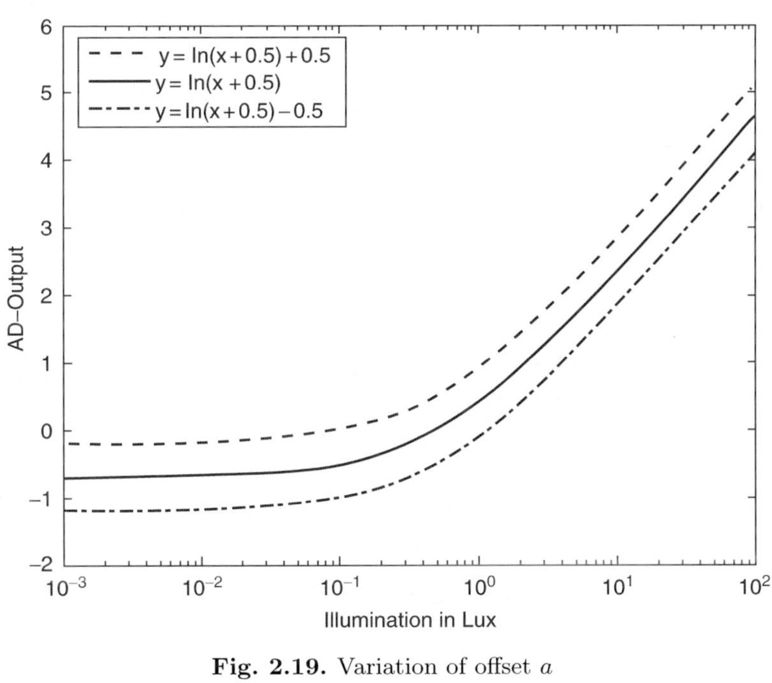

Fig. 2.19. Variation of offset a

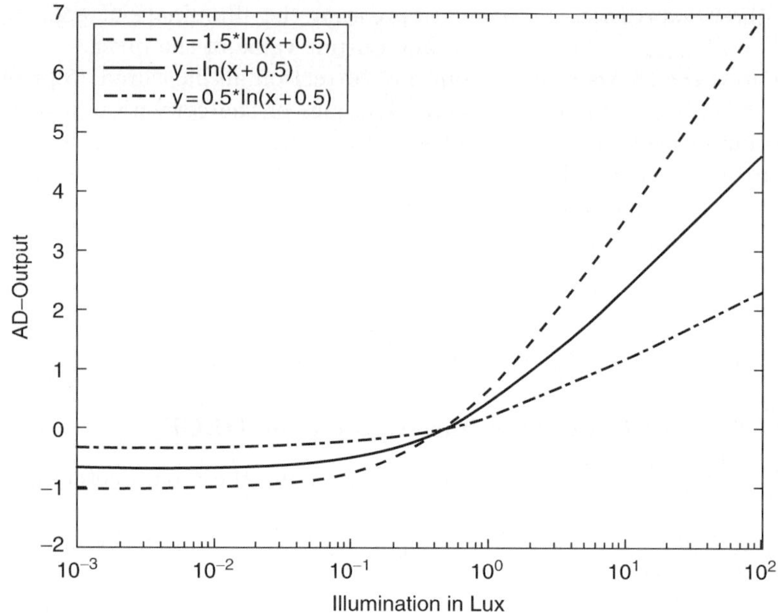

Fig. 2.20. Variation of gain b

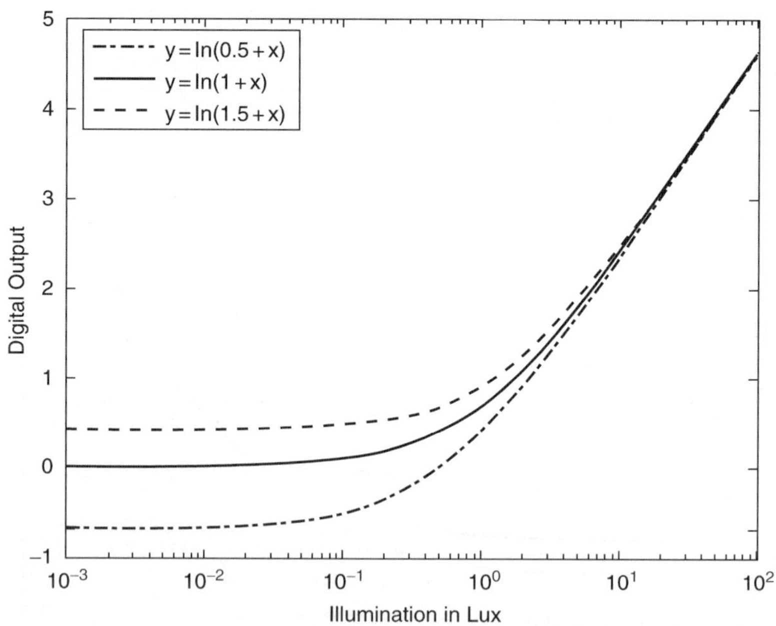

Fig. 2.21. Variation of the 3dB-point c

2.4.4 Presentation of Three Correction Algorithms

In the literature three correction algorithms are named which either consider only a variation of parameter a (1-parameter correction), a correction in the offset and gain (2-parameter correction) or of all three parameters (3-parameter correction) [4]. The following set of equations defines the correction values for each algorithm:

$$
\begin{aligned}
y_{1-\text{Parameter}} &= y_i - a_i + \bar{a}, \\
y_{2-\text{Parameter}} &= \bar{b} \cdot \frac{y_i - a_i}{b_i} + \bar{a}, \\
y_{3-\text{Parameter}} &= \bar{b} \cdot ln\left(\mathrm{e}^{\frac{(y_i - a_i)}{b_i}} - c_i + \bar{c} \right) + \bar{a}.
\end{aligned}
\tag{2.16}
$$

Figure. 2.22 compares the deviation of the 1-, 2-, and 3-parameter correction algorithms. The 1-parameter correction represents the worst case scenario, as it does not consider the gain and 3 dB-point variation. The 2-parameter correction also has a high deviation in the dark; under bright conditions, however, it delivers as good results as the 3-parameter correction. This means that the best correction is the 3-parameter correction, which considers all three parameters.

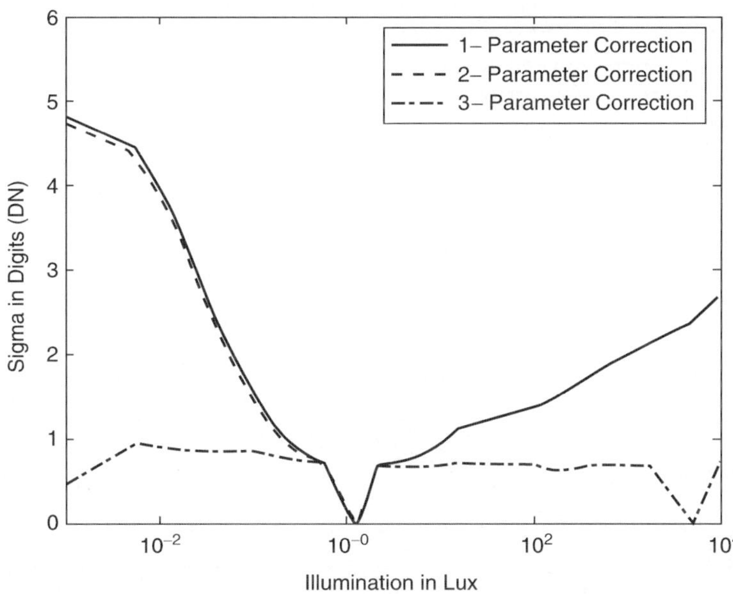

Fig. 2.22. Comparison of 1-, 2- and 3-parameter correction algorithm

2.4.5 New Parameterized Correction Algorithm

As the correction algorithm for the 3-parameter method requires exponential and logarithmic calculus (see (2.16)), it cannot be implemented in cheap hardware for everyday use. In order to solve this problem, the newly developed algorithm approximates the logarithmic curve by three lines:

Simple methods use two lines for approximation: line L1 in dark and line L3 in bright conditions. Line L1 represents a simple offset correction in the dark, as it does not consider a gain factor. Under bright conditions line L3 approximates the logarithmic curve (Fig. 2.23). It becomes apparent, however, that the error of this simplification is high in the area around the 3 dB-point. In order to solve this problem, we use a third line L2 which is a secant to the logarithmic curve and produces half the gain L3 does. Simulations showed that the secant delivers better results than the tangent through the 3 dB-point. Line L2 is used for correction between the illumination $c < x < 4 * c$. All three lines can be calculated with the help of parameters a, b and c:

$$\begin{aligned} y_{L1,i} &= a_i + b_i \cdot \ln(c_i) = lin_i \\ y_{L2,i} &= a2_i + b_i \cdot \ln(x) \text{ with } a2_i = a_i + 0.5 \cdot b_i \cdot \ln(c_i) + b_i \\ y_{L3,i} &= a_i + b_i \cdot \ln(x). \end{aligned} \tag{2.17}$$

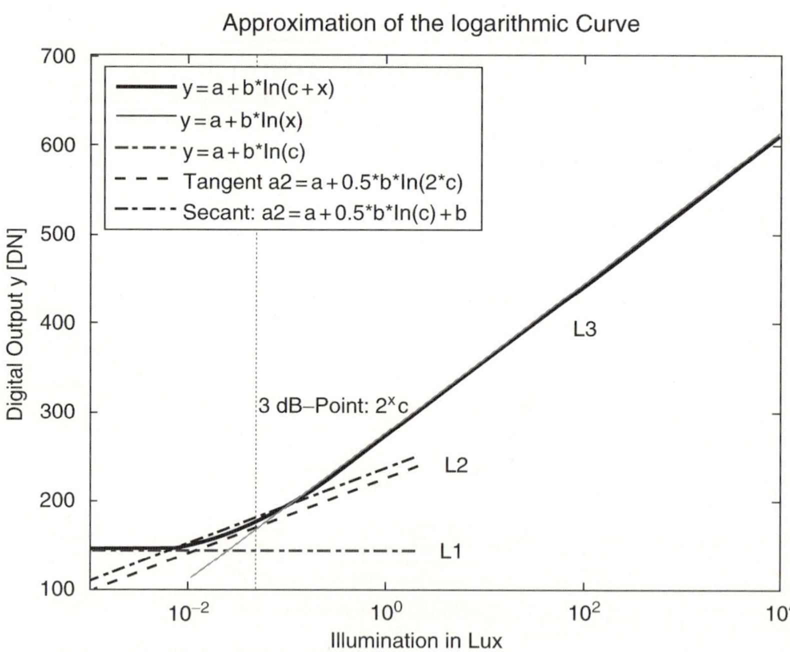

Fig. 2.23. Approximation of the logarithmic characteristic

As described before the parameters for each pixel are defined by measurements under at least three different illuminations. The succeeding set of equations describes how the raw pixel data is corrected during runtime. The presented algorithm does not need any logarithmic or exponential calculus and is, thus, easy to implement into hardware

$y_{\mathrm{raw},}\ i \leq$ thresh1:

$$y_{cor,\,i} = y_{\mathrm{raw,i}} - (a_i + b_i \cdot c_i) + \overline{lin} \tag{2.18}$$

thresh1$\leq y_{\mathrm{raw,i}} \leq$thresh2:

$$y_{cor,\,i} = \overline{a2} + \overline{b2} \cdot \frac{y_{raw,\,i} - (a_i + 0.5 \cdot b_i \cdot c_i + b_i)}{0.5 \cdot b_i}$$

thresh2$\leq y_{\mathrm{raw},\,i}$:

$$y_{cor,\,i} = \bar{a} + \bar{b} \cdot \frac{y_{raw,\,i} - a_i}{b_i}$$

Figure 2.24 compares the new parameterized algorithm ((2.18)) with the logarithmic 3-point correction (see (2.16)). The analyzed image sensor without correction has a 10-bit resolution and a 22-digit deviation. Thus, the two algorithms deliver very good results and reduce the FPN by a factor 17. In this case, both methods use three measuring points in the dark for further

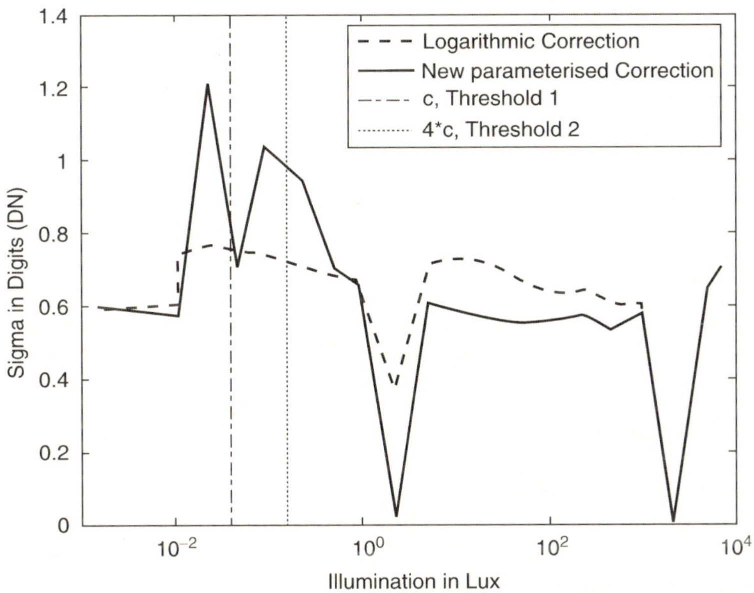

Fig. 2.24. Comparison of logarithmic correction with new parameterized algorithm

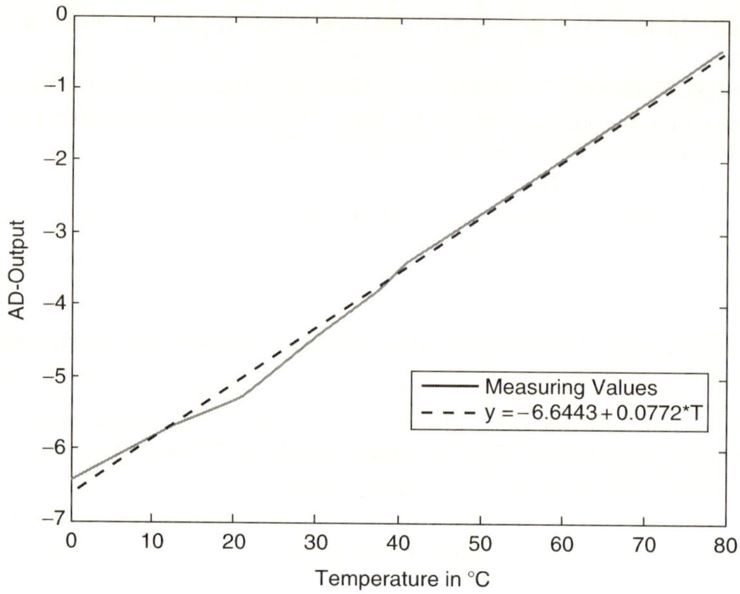

Fig. 2.27. Temperature dependence of offset *a*

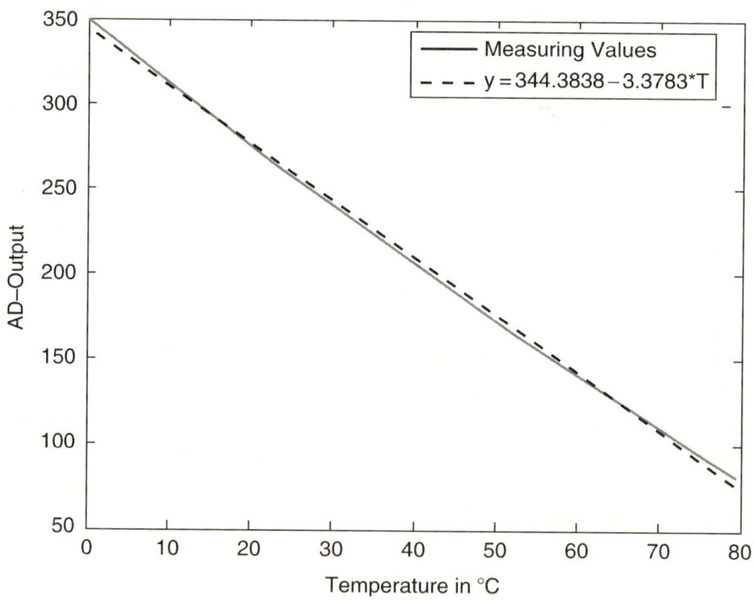

Fig. 2.28. Temperature dependence of gain *b*

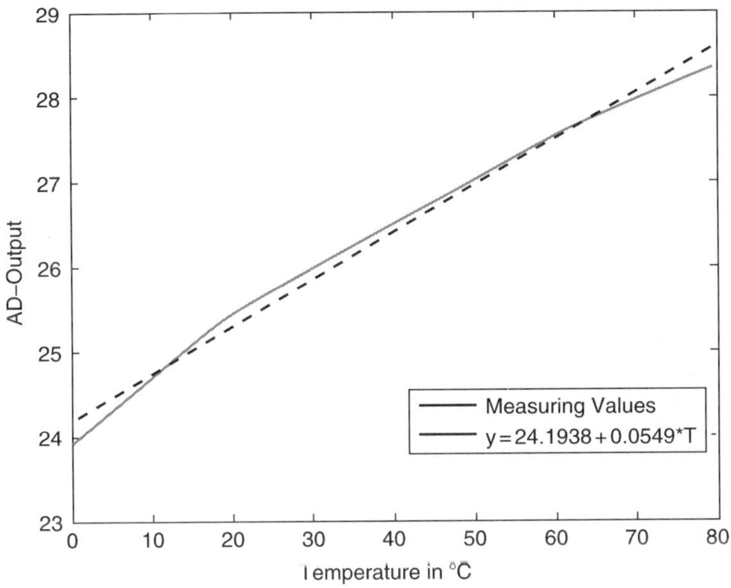

Fig. 2.29. Temperature dependence of the 3 dB-point c

$$\begin{aligned} b_i(T_{^\circ C}) &= b_{i,0} + \beta_i T_{^\circ C}. \\ c_i'(T_{^\circ C}) &= c_{i,0}' + \gamma_i T_{^\circ C}. \end{aligned} \tag{2.20}$$

Including the parameter temperature dependence leads to a doubling in the amount of parameters to be saved. However, compared to other approaches which save a set of parameters ($a_{i,t}$, $b_{i,t}$ and $c_{i,t}$) for different temperatures t individually with only six parameters to cover the whole temperature range is very good.

Figure 2.30 shows how the deviation in illumination and temperature of the new algorithm including temperature varies. First of all, it should be mentioned that for the first general analysis it is sufficient to take the same set of data to calculate the correction data as well as implement the correction itself. Furthermore, the resolution again is 10 bit and the deviation without correction is roughly 22 digits across the whole temperature range. Thus, the maximum peak at 1.8 is optimized by a factor 12. The figure also shows that the deviation increases with temperature and it proves that our new approach works correctly at different temperatures. Under bright conditions, a slight increase in the deviation becomes apparent. This is caused by an optimization to 40 bit per pixel. In this case, parameter β is not saved per pixel but rather per column. When using 48 bit the deviation is constant under bright conditions. Fixed-point calculation is needed, where all six parameters are 8 bits long. Again, the same hardware is used: XILINX-FPGA XCV400E with 8 and 16 bit storage (see Table 2.5).

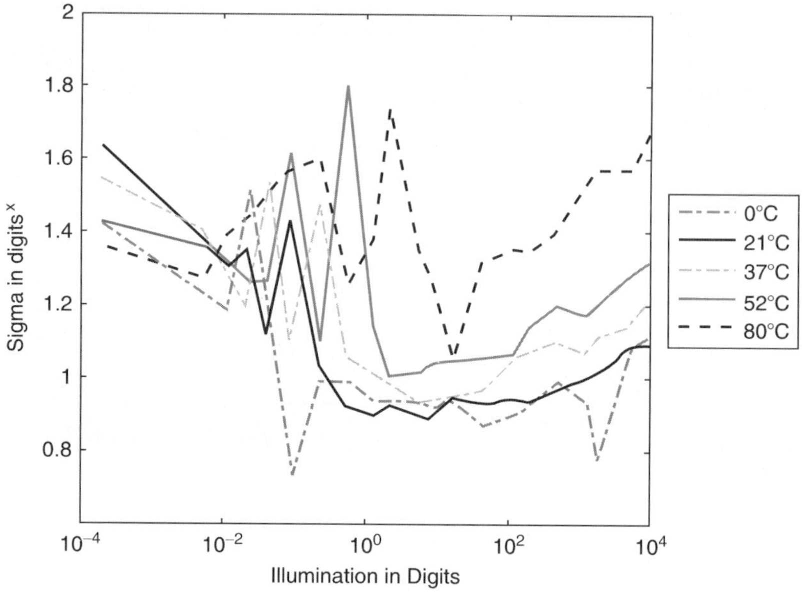

Fig. 2.30. Deviation over illumination and temperature – correction results with the algorithm including temperature [5]

Table 2.5. Parameter resolution for method including temperature compensation

Parameter	Resolution
a_0	2
b_0	2^{-3}
c_0	2^{-4}
α	2^{-5}
β	2^{-10}
γ	2^{-10}

Ultimately, this chapter introduced a new correction method which needs only one set of data of 40/48 bit to correct fixed-pattern noise across the entire temperature range. This method leads to better results and, for that reason, extends the application field of HDRC image sensors.

2.4.8 Correction Procedure and Runtime

Finding a suitable good fixed-pattern correction algorithm is only the first step towards high quality images. After the production of HDR imagers, their specific fixed-pattern correction parameter set must be determined. For this we first measure the sensor characteristic by using a phlox-plate followed by the PC calculating the correction parameters according to the previously presented algorithm using (2.14). The exponential calculus in these equations

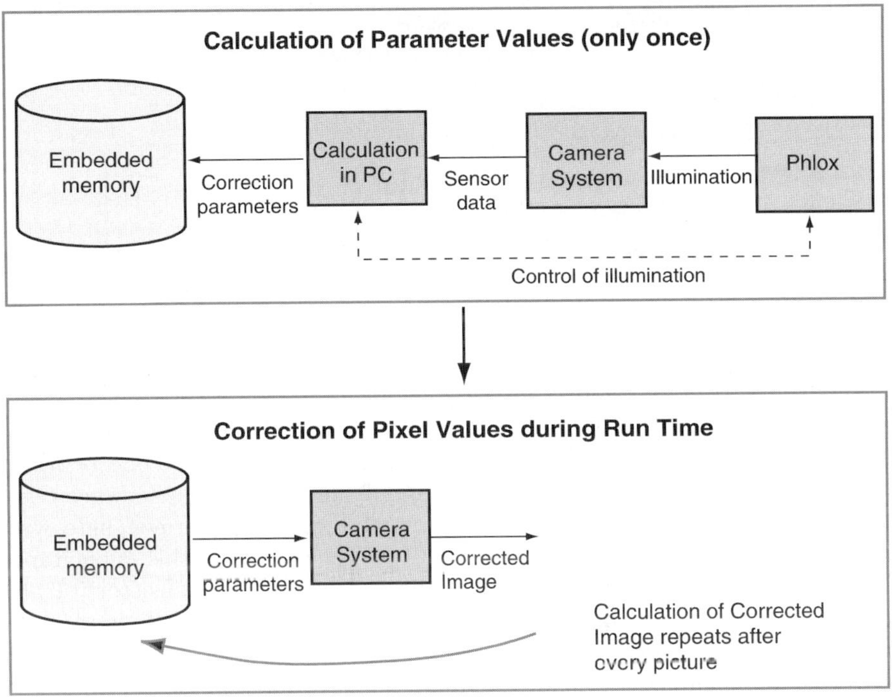

Fig. 2.31. Process flow of fixed-pattern correction

is not problematic as it is performed by a fast computer. Finally, this data is transferred to the embedded memory of the sensor.

During runtime, the digital correction unit reads out the raw data of the sensor and corrects it. The hardware first evaluates the raw value with the threshold and then decides on (2.18). Figure 2.31 visualizes this process.

2.4.9 Summary

The presented method regards all physical pixel parameters as the offset, the gain and the dark current of the pixel. The results show that the algorithm improves image quality greatly compared to the 1-point-offset correction. A comparison at high temperatures shows that the image quality can be improved by using temperature compensation. The strategy and the realization of this method are based on the effectiveness and speed of digital storage and processing so that a correction within the camera head is possible in realtime with 32 frames s^{-1} at VGA-resolution (300,000 pixels).

2.5 HDRC Dynamic Performance

As a continuous photocurrent monitor with logarithmic opto-electronic conversion function, the HDRC sensor has an inherently simple functionality. The

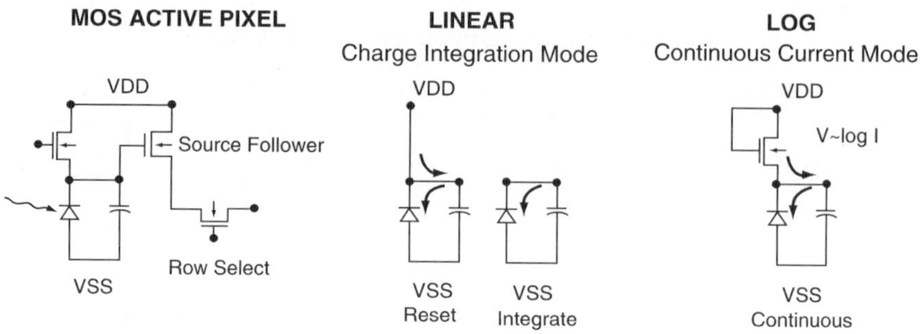

Fig. 2.32. HDRC pixel with reset including key capacitance

output voltage of an HDRC pixel continuously tracks changes in the photocurrent. Naturally, the node voltage Vlog tracks the changes in the photocurrent depending on the charging and discharging of the log-node capacitance C.

In Fig. 2.32, the HDRC pixel diagram of Fig. 2.7 is redrawn here with its significant capacitance. In our HDRC reference sensor, this capacitance is typically 2.5 fF. This capacitance is charged by the log-transistor T1 and discharged by the photodiode current. In the stationary state, the node voltage Vlog is dictated by the gate-source voltage VGS, which is consistent with the photodiode current according to (3.1). This means that for low light levels Vlog is high, typically 2.8 Volt in a 3.3 Volt sensor and it is low, about 2.4 Volt, for high illuminance values. If the illumination of a pixel changes abruptly from dark to bright, Vlog is high, VGS is low and the transistor conductance is very small. The pixel capacitance is discharged by the increased photocurrent and as Vlog decreases, VGS is increased and a competing pull-up current is generated by the transistor T1. Nevertheless, in this case the transition is mostly determined by a discharge of the pixel capacitance with a constant current source. For a transition from 10 mlx to 10 lx the settling time for Vlog is much less than one frame time.

For a step-function change from bright to dark, the pixel capacitor has to be charged to a higher voltage. Initially, the gate-source voltage is fairly large and, consequently, sufficient pull-up current is delivered by transistor T1. However, as Vlog rises, VGS decreases and, consequently, the pull-up current decreases exponentially resulting in an ever slower rise of Vlog.

Figure 2.33 shows Vlog versus time for this case and it is evident that the pixel does not obtain its stationary dark-level voltage within one frame. Practically, a laterally moving bright light source leaves behind a tail of grey levels like a comet tail. This phenomenon has been called HDRC image lag.

For this transition from bright to dark, the fact that the HDRC pixel tracks differences continuously with some memory of the previous frame causes this effect.

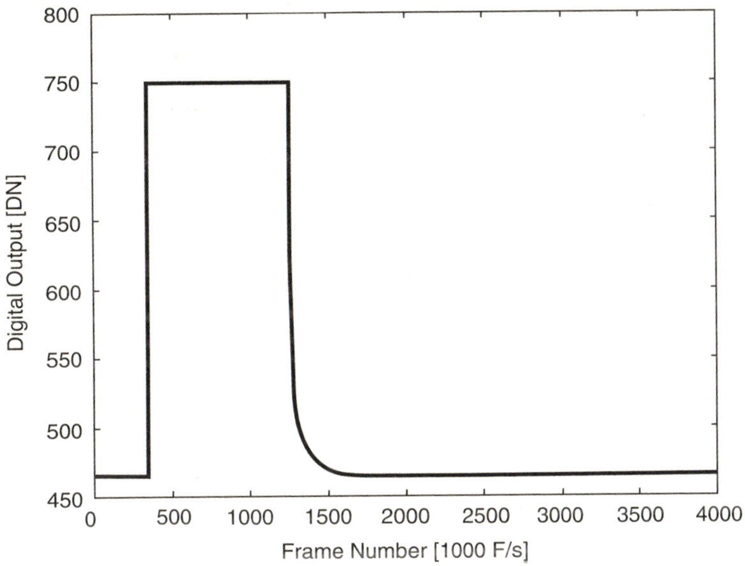

Fig. 2.33. HDRC pixel voltage Vlog after a step-function change from 1 klx to 10 mlx

In order to erase this memory, Vlog has to be reset for each frame to a reference level. The log transistor T1 can serve for this purpose. For this function, T1 is operated as an active transistor with properly controlled gate and drain voltages. If the gate voltage VGG is sufficiently high compared with the drain voltage SVDD, the difference being at least the threshold voltage VT, Vlog will be forced to SVDD. This "dark" reset was proposed basically in 1997 [6]. The most effective mechanism is outlined in Fig. 2.34.

At the end of a frame, the Vlog voltages are distributed from dark to bright as shown in the figure, obviously with VGG = SVDD = VDD as in the classical continous HDRC mode. Reset is initiated by raising VGG to VGGH and lowering SVDD to a level SVDDL, which is slightly higher than the dark level of Vlog. All pixels are effectively black at the end of reset. Now VGG and VDD return to the standard HDRC level SVDD. This condition effectively turns off the transistor, and all pixel nodes Vlog are now discharged by their individual photocurrent. Brightly illuminated pixels with large photocurrents will see a rapid decrease of Vlog, turning on the log-transistor T1 and quickly establishing the stationary bright signal voltage Vlog identical to the continuous HDRC mode. Dark pixels with low photocurrent will be discharged slowly with a linear response dictated by the pixel capacitance C until, for sufficiently large integration time, Vlog (t) has decreased to values where the log transistor T1 begins to provide a competing pull-up current so that Vlog settles to a value closer to the log-response value of the continuous mode of operation.

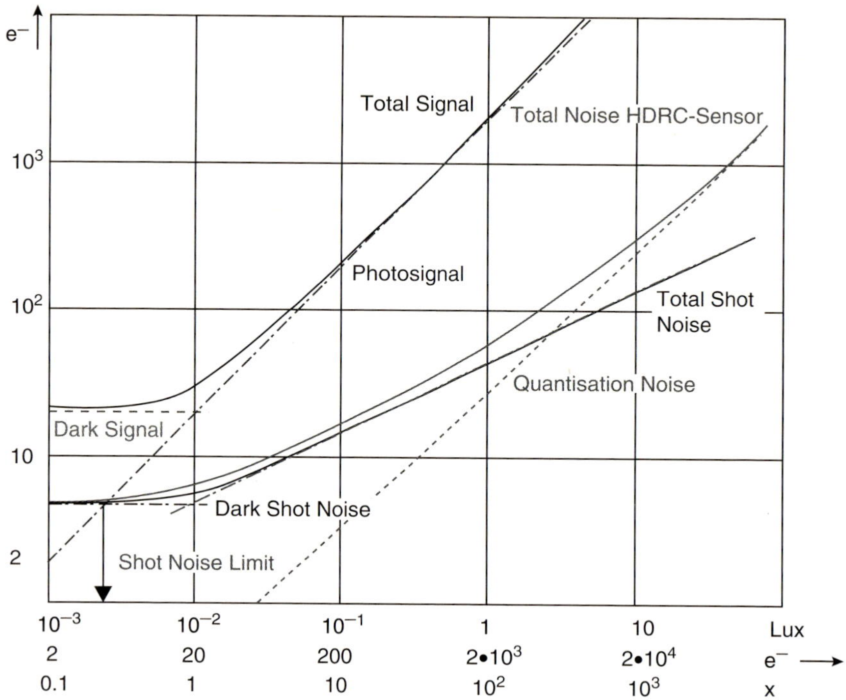

Fig. 3.5. Flat-field noise HDRC sensor (40 fps)

In Fig. 3.4 we have shown the signal-to-noise ratios of our reference CMOS sensor both in the linear reset mode and in the high-dynamic-range continuous mode. We notice that the HDRC sensor has the superior flat-field signal-to-noise ratio over an extremely wide dynamic range starting out with a better ratio at very low light levels achieving a maximum already at fairly low light levels and maintaining this ratio over six orders of magnitude. The linear integrating sensor is inferior at low levels because of reset and quantization noise. It is superior near its white saturation but of course at the price of white saturation and a useful dynamic range of less than 60 dB.

The signal-to-noise characteristic of the HDRC sensor shown in Figs. 3.4 and 3.5 is correct only for quasistationary illumination. For rapid changes of illumination including low illumination levels, the HDRC sensor has to be operated with a dark or white reset as described in Sect. 2.4. In this case, we have to include the reset noise. At low light levels, the HDRC sensor with dark reset due to its higher contrast sensitivity in the 1 lx region exceeds the continuous HDRC mode. It still offers the very high-dynamic-range and is attractive because of the elimination of the image lag. At lower light levels, because of its somewhat reduced contrast sensitivity, the white reset mode of the HDRC sensor is inferior to the continuous mode, but again, it is free from image lag and offers the same very high-dynamic-range.

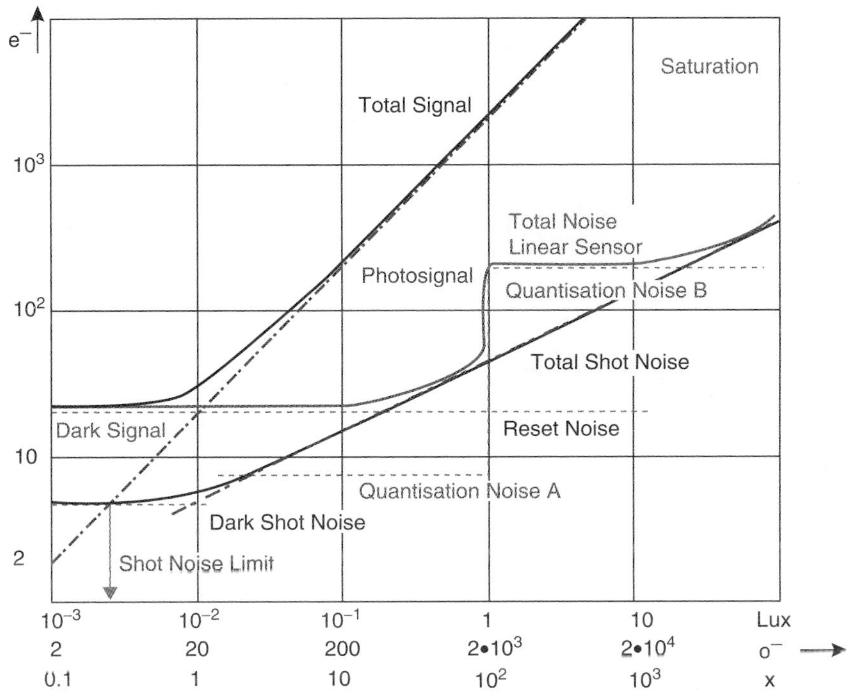

Fig. 3.6. Flat-field noise piecewise linear APS sensor

In the linear integrating mode of CMOS pixels, the dynamic range can be extended by switching to a smaller gain above a certain threshold. The effect of this technique on the flat-field noise is shown in Fig. 3.6, where this threshold is set at 2,000 e and where the gain drops by a factor of 30. The S/N ratio for this technique is also included in Fig. 3.4.

It should be noted that the inverse of the flat-field signal-to-noise ratio S/N in Fig. 3.4 gives us the detectable contrast for 40 frames s^{-1}.

Because we have applied one flat-field noise model and one common reference pixel, the favorable results for the HDRC sensor make it very attractive. At the same time, we have to keep in mind that these noise results are all based on the assumption that we have performed a digital fixed-pattern correction with a standard deviation of 1 DN over the full dynamic range following the methods outlined in Sects. 2.2 and 2.3.

References

1. Janesick JR (2003) Introduction to CCD and CMOS imaging sensors and applications. SPIE Education Services SC504 San Diego, CA USA

4

High-Dynamic-Range Contrast and Color Management

Bernd Hoefflinger

By now we have learned that HDRC sensors faithfully acquire scenes with widest dynamic range and with very good sensitivity and uniformity. The sensor signals we now have available are the prerequisite for a high-performance electronic-vision system, which should accomplish two tasks:

1. Present and display the scene informative and pleasing to our eyes.
2. Deliver optimal input signals to machine-vision systems, which build on the unique characteristics of the human vision system (HVS) combined with the benefits of digital processing power.

For both tasks, the HVS provides guidance on many counts. We concentrate here on three:

1. Contrast sensitivity: How do we detect small differences anywhere in any scene?
2. Discount the illuminant, which can be extremely variable, spotty and blinding, obscuring the objects we want to see.
3. Provide correct and constant colors under different and changing illuminations both for our critical perception of colors and for object recognition and classification in machine-vision systems.

We recapitulate that the quality of an imager lies in its capability to help us perceive detail and differences in the optical signals, which we acquire. In sensing the optical intensity I, what is the minimum just noticeable difference (JND) ΔI? It was Weber's finding in the 19th century that, for many human sensing functions, the ratio $\Delta I/I$ is constant. This law, extended over the wide dynamic range of the eye results in the famous logarithmic relationship for the detected signal S:

$$S = K_0 \ln I. \tag{4.1}$$

The constant K is the contrast sensitivity, which we saw in Chap. 2 is about 1% for the human eye and it is about 1% for an HDRC sensor with 140 dB dynamic range and a 10 bit digital output. The sensitivity decreases at very low intensities so that the full OECF is

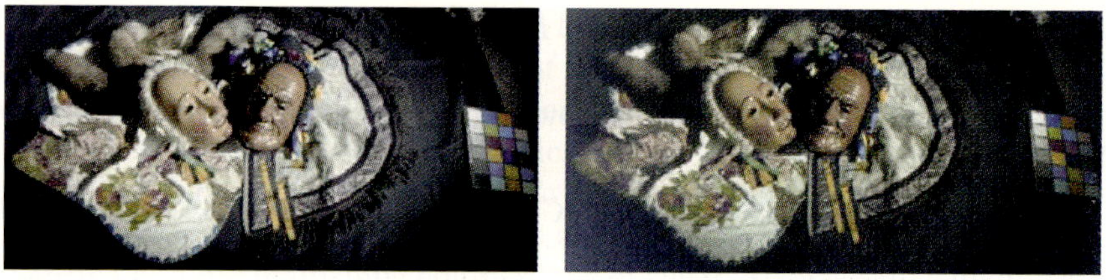

Fig. 4.3. HDRC recordings of masks with two apertures after log luminance offset of the pixel data. *Left*: 1:1.8. *Right*: 1:8

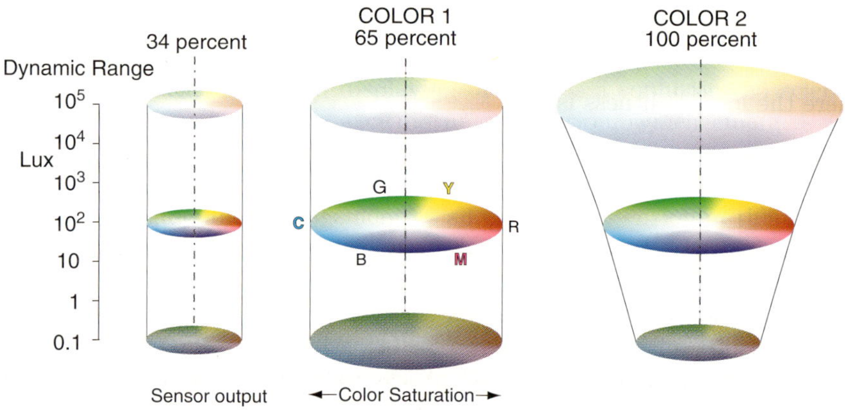

Fig. 4.4. The cones of chromaticity vs. lightness

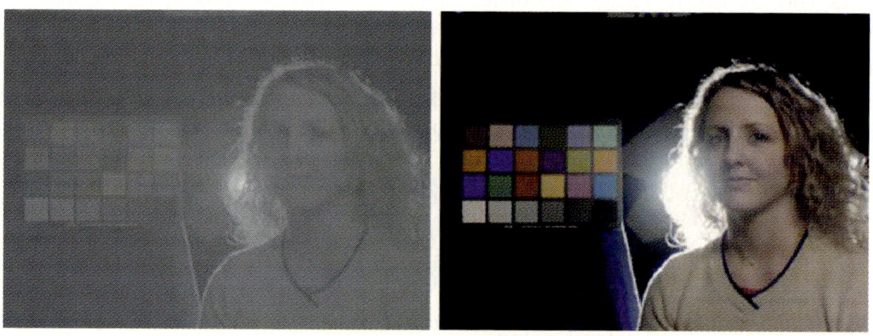

Fig. 4.5. HDRC color image before and after enhanced color saturation

Table 4.1. Steps to ward displaying log HDR color images

1. For the R, G, B pixels, record the output voltages V_R, V_G, V_B of the log-converting HDRC sensor and convert to digital numbers (DN).
   (a) With a Gray opaque diffuser filter, shift the centers of the recorded RGB histograms to coincide = white balance.
   (b) Remove filter, and for each captured frame, get Min and Max of the frame histogram.
2. Expand and shift all DN to fully utilize the lightness range L^*_W.
3. Depending on the RGB mosaic topology, perform center-surround operations to obtain the tricolor C_R, C_G, C_B for each pixel.
4. Perform color enhancement operations to obtain C^*_R, C^*_G, C^*_B per pixel.

It has been suggested for a linear OECF to amplify the differences (R–L, G–L, B–L). However, for a linear OECF, more precisely, the ratios $R/L, G/L, B/L$ should be amplified. The difference is only justified and appropriate for a logarithmic OECF.

The absolute and relative magnitudes of the color amplification factors α_R, α_G, and α_B have to be determined for a given HDRC sensor and camera. Typical results for the VGAx sensor are $\alpha_R = \alpha_G = \alpha_B = 2.0$. Amplification may have to increase with lightness dynamic range to maintain color saturation. As a result, the lightness – color cone of Fig. 4.4 is obtained for a HDRC color camera.

Figure 4.5 shows a typical HDRC color image before and after saturation enhancement.

To summarize, the steps for a densitometric HDRC color image are listed in Table 4.1.

In this chapter we have limited HDR contrast and color management to simple linear global operations for the optimal display of recorded HDRC images, which can be executed easily frame-by-frame in real time. More advanced and particularly perception-motivated HDR video tone mapping and encoding will be treated in Chaps. 11 and 12.

References

1. Giorgianni EJ, Madden TE (1997) Digital color management. Addison-Wesley, Reading
2. Johnson GM and Fairchild MD (2003) Visual psychophysics and color appearance. In: Digital color imaging handbook, Sharma G ed., Chap. 2
3. Wyszecki W and Stiles WS (2000) Color science: concepts and methods. Wiley, New York

5

HDR Video Cameras

Markus Strobel and Volker Gengenbach

5.1 Introduction

In this chapter, we describe various cameras built with HDRC sensors. One common property of these cameras is the modular system design which separates the functionality in a sensor board, a controller board, and an interface board for different communication standards, e.g., CameraLink^{TM}, NTSC/PAL, or proprietary data transmission.

The optical properties of the cameras are determined by the HDRC sensor itself, so a brief overview of the actual HDRC sensors is given in this introduction.

Table 5.1 gives the technical specifications of the HDRC VGAy sensor with rolling shutter and the HDRC-G sensor with global shutter. The monochrome version of the sensors has a wide spectral sensitivity range from 400 to 950 nm, the color version has an additional RGB-Bayer color mosaic filter. As the monochrome sensor is sensitive to infrared radiation, it is the sensor of choice for automotive night vision applications with active near infrared (NIR) illumination.

The HDRC VGAy sensor with on-chip A/D-converter and rolling shutter is explained more detailed below. It is well suited for high performance cameras, monochrome or color, as well as for highly miniaturized camera modules. The block diagram is shown in Fig. 5.1.

The HDRC VGAy is a high-dynamic-range CMOS imager with 10 bit digital video output, which compresses a wide range of light intensity instantaneously and within each pixel with a logarithmic response like our eyes, but exceeding their spontaneous range by orders of magnitude. The ultra-high-dynamic-range, which is larger than 160 dB (>8 decades of light intensity) leads to saturation-free images. This allows imaging without the need for exposure control like aperture, shutter, or integration time control. Pixels can be read non-destructively at high-speed, which results in very high frame rates, e.g., more than 4 k subframes s^{-1} consisting of 3,000 s^{-1} pixels. Additionally,

Fig. 5.3. HDRC CamCube, miniaturized camera module

Fig. 5.4. Exploded view of the HDRC CamCube

5.2.1 Features of the HDRC CamCube

- HDRC VGA sensor with 640×480 pixels
- >140 dB dynamic range
- 10 bit digital output
- 30 fps at full-resolution
- 10 MHz pixel rate
- User defined subframe readout (AOI)
- MMRs of the sensor can be set by a serial interface
- Very compact module, only $25 \times 25 \times 25\,\mathrm{mm}^3$ (H, W, L)
- Standard NF-mount lens with adapters to C-mount or 12×0.5 mm lenses

5.2.2 Assembly Techniques

Figure 5.4 shows the exploded view of the HDRC CamCube with the sensor board (to the right), followed by the board-to-board connection spacer, the

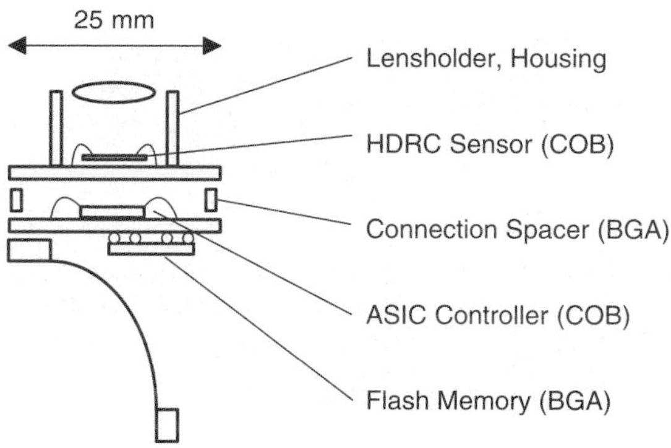

Fig. 5.5. Assembly of the HDRC CamCube (schematic)

controller board and output via a 32 pin flat connection cable which can have a length of up to 20 cm.

For the assembly of the module several mounting techniques like chip on board (COB), ball grid array (BGA), and surface mount technology (SMT) were used in order to shrink the dimensions of the printed circuit boards. See Fig. 5.5 with the schematic assembly and the main components HDRC sensor, ASIC controller, and flash memory.

- *COB*: Direct chip mounting for the HDRC sensor and the ASIC controller on the circuit board. The bondwires of the chips were covered with a globe top for sealing.
- *BGA*: The flash memory and the board-to-board spacer are mounted by solder bumps in a BGA technology.
- *SMT*: SMT for all other circuits like the oscillator, the resistors, and capacitors.

5.2.3 System Design

Figure 5.6 gives an overview of the functional blocks and the signal chain of the HDRC CamCube from the pixel with the optoelectronic conversion of the incident light, the multiplexed readout of the pixels, the A/D-conversion of the image data, the fixed-pattern correction (FPC) in the controller with the correction data from the flash memory.

The HDRC VGA sensor used here is a prior version and is similar to the HDRC VGAy sensor described in Sect. 5.1. Due to the small size of the module the use of an on-chip A/D-converter is necessary. The 10 bit output data are fed into the controller which addresses the FPC memory, here a 512 k $\times$ 8 bit flash memory with pixel rate access time. To reduce the size and power consumption of the camera module the controller was realized as an ASIC instead of using an FPGA.

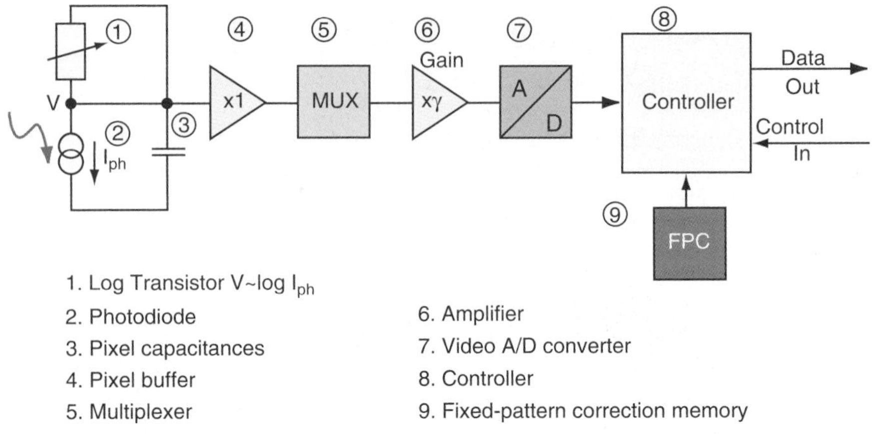

Fig. 5.6. Block diagram of HDRC CamCube

5.2.4 Application Example

Figure 5.7 shows an application example of the HDRC CamCube mounted in an interior mirror of a car and Fig. 5.8 shows the interfacing to the Camera Link™ standard.

5.3 HDRC Camera Front-End

The HDRC Camera Front-End [1], Fig. 5.9, is intended to serve as a reference design for a quick integration of HDRC CMOS sensors in digital vision systems. Physical interfacing to any transmission board with standards like RS422, Camera Link™ [2], 1394 Fire Wire, or USB2.0 is easy to accomplish by the generic digital video and synchronization signals of the interface. The

Fig. 5.7. HDRC CamCubes integrated in an interior mirror of a car looking on the road through the windshield

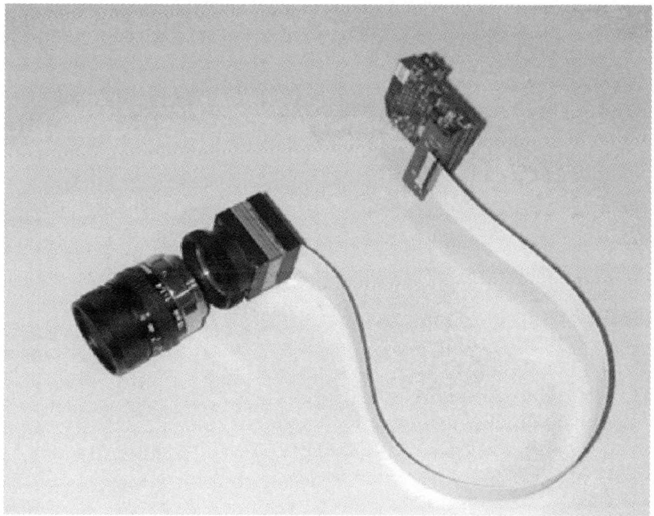

Fig. 5.8. HDRC CamCube with a Camera Link^{TM} interface board

Fig. 5.9. HDRC Camera Front-End

interface is open for active controlling the HDRC sensor's registers via a bidirectional serial interface, e.g., accessed by a microcontroller or a digital signal processor. The HDRC Camera Front-End as well as the interface is designed for highest demands in image capturing and transmission. The system concept allows smart solutions with minimum hardware requirements.

An integrated lens mount adapter provides direct adaptation of special HDRC lenses but, at the same time, offers an interface to standard C-mount lenses. The dimensions of the HDRC Camera Front-End with the sensor and controller board are depicted in Figs. 5.10 and 5.11, respectively. The optical center is defined by the sensor's center of the active pixel array.

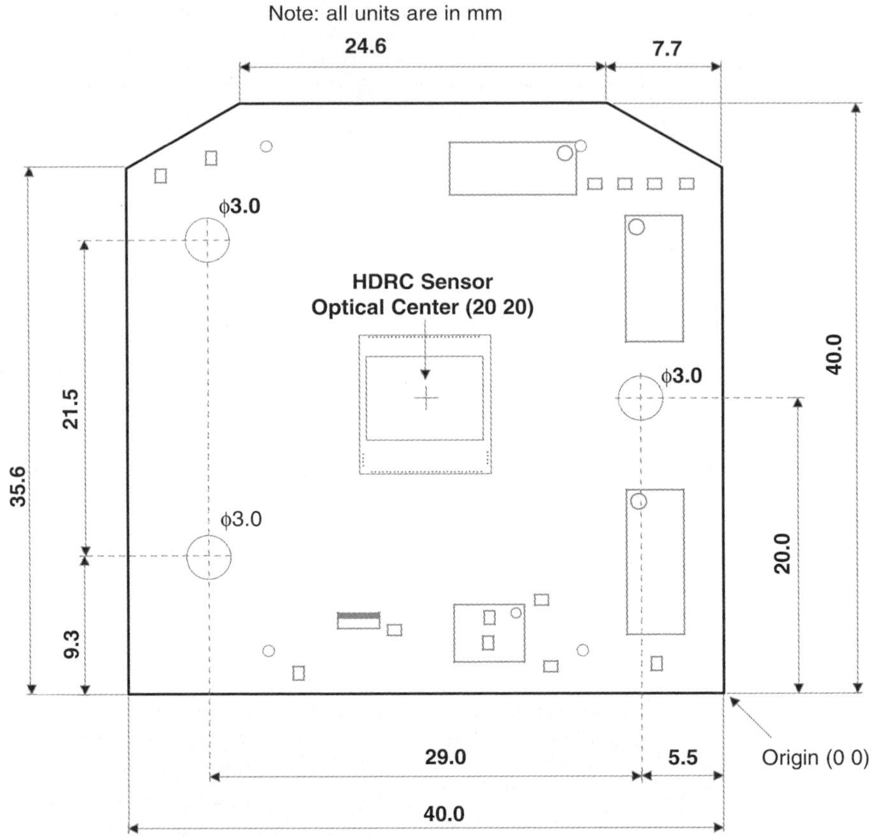

Fig. 5.10. Front side of the HDRC Camera Front-End with dimensions

A flexible 24 bit parallel data bus with bit assignments of 1 × 8, 1 × 10, 1 × 12, 2 × 10, 2 × 12, 3 × 8, or up to 1 × 24 bit allows adaptation of future HDRC imager generations with high resolution (Megapixel or HDTV) and high-speed digital output. The pinout of the module connector is shown in Fig. 5.12, and Table 5.2 gives the pin definition of the digital signals and power supply.

The signals in Table 5.2 can be grouped according to their function into video data, frame synchronization, serial control, and system signals as well as power supply:

- Video data DIO23 to DIO0. DIO23 to DIO12 for a 12 bit, DIO23 to DIO14 for a 10 bit single digital output are assigned, respectively.
- Frame synchronization signals. /LEN and /FEN define the horizontal and vertical frame synchronization, respectively. PCLK defines the pixel rate.

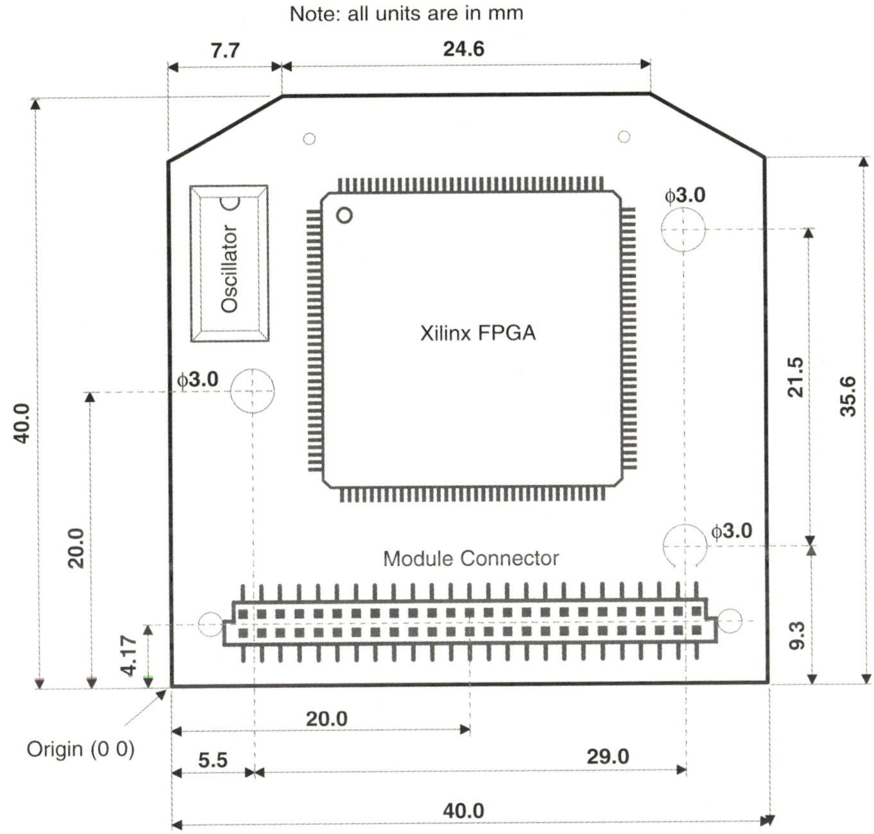

Fig. 5.11. Back side of the HDRC Camera Front-End with dimensions

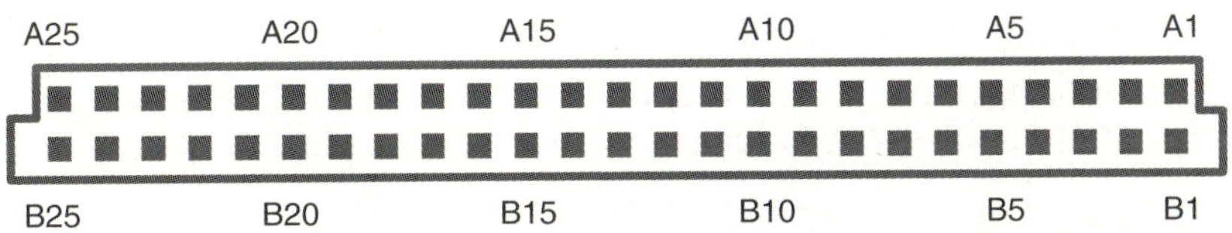

Fig. 5.12. Pinout of the module connector (type: ERNI SMC-B-50F, female)

- Serial control interface. With the signals SDIN, SCLK, SLD, and SOUT, a SPI slave interface of the very common serial peripheral interface (SPI) standard is used.
- System signals. TRIGGER (input), /RESET, and oscillator OSC (input or output) can be used for triggering on single events or pixel synchronous readout of multiple HDRC Camera Front-Ends, e.g., for stereo applications.

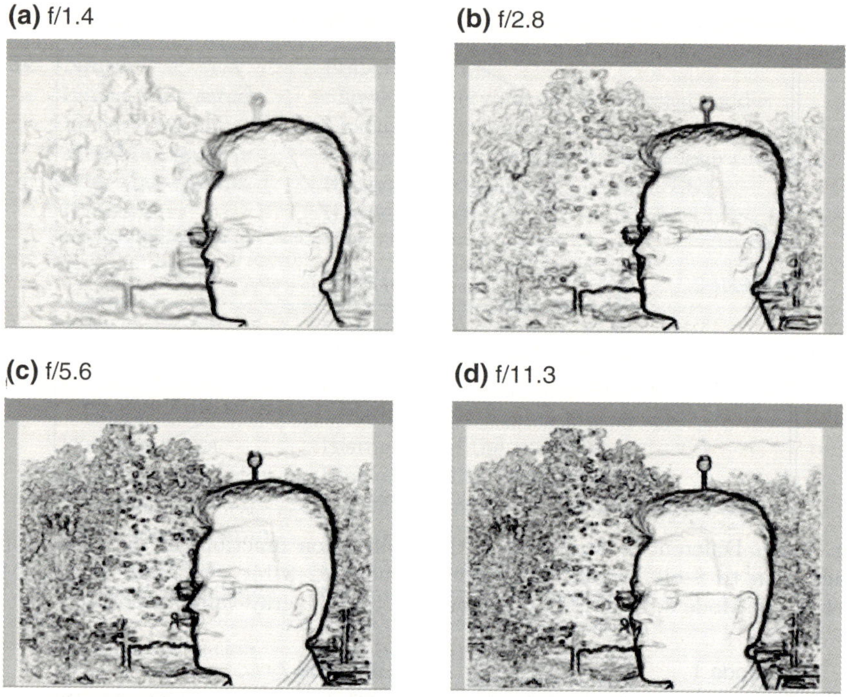

Fig. 5.22. Sequence of edge images at different f-stops of the lens in processing Mode B of the camera: (**a**) $f/1.4$; (**b**) $f/2.8$; (**c**) $f/5.6$; (**d**) $f/11.3$. *Note*: for better visibility the gray levels of the edge images are inverted

- With decreasing f-stop the depth of focus of the lens increases, therefore edges in the trees appear more and more.
- As a result we can see that with the HDRC sensor a robust edge detection without adapting thresholds over the illumination is possible.

5.6 HDR Video-Based Aircraft Docking Guidance

5.6.1 Summary

One important task for the pilot of a large passenger aircraft is driving into the exact parking position with precision. Due to a very restricted visual field the pilot needs help from outside the aircraft to successfully solve this task. An automatic docking guidance system helps the pilot by displaying navigational information on a large panel and ensures precise guidance of the aircraft over the final few meters to its correct docking position. Such a docking guidance system can be built using a video camera combined with a model-based image-processing system to detect the aircraft and sense its position.

The camera has to fulfill extreme requirements, which can be handled by an application-specific HDR video camera.

5.6.2 Introduction

Boarding bridges at modern airports allow passengers to get on or off the aircraft in a very comfortable way. These bridges are maneuverable but only within a limited area. In order to guarantee smooth operation of passenger boarding bridges, the aircrafts have to be parked as close to their final parking position as possible with all types of aircrafts. The pilot's visual field is very limited and does not allow him to park the aircraft at the docking position with sufficient precision. He needs guidance from outside the aircraft. This can be given by a person, the so-called "marshaller," who guides the pilot to the parking position by signaling with his hands or a light stick.

The increasing number of daily aircraft movements and the widespread concentration of take-off and landing schedules during peak times force airport administration at large airports to fully automate the docking process. They install automatic docking guidance systems which guide the pilot with the help of a large display panel. On the display panel the pilot can continuously read the distance of the aircraft to the docking position and the lateral deviation of the aircraft from the guidance line. In order to calculate the values shown on the pilot display, a sensor is needed which measures the position of the approaching airplane with high precision, low latency as well as a high refresh rate. The described visual docking guidance system (VDGS) [4] distributed by Honeywell Airport Systems GmbH [5] having already been installed hundreds of times at airports throughout Europe and Asia uses a HDR video camera in combination with an image processing unit (see Fig. 5.23).

The demands on automatic docking systems are very high: the systems have to function with utmost reliability and precision regardless of the time of day or weather conditions as long as aircrafts proceed to take-off and land.

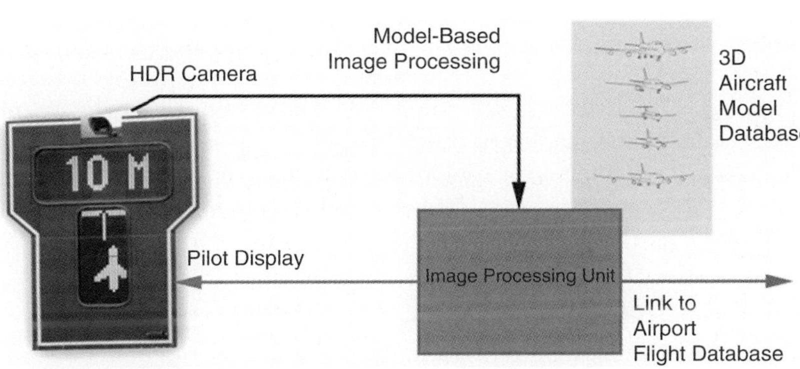

Fig. 5.23. Structure of the video-based aircraft docking guidance system

In order to allow the pilot the precise navigation to the parking position, the aircraft's position has to be displayed with very short latency. The sensor system should be able to recognize the arriving aircraft to avoid irritations of the system by other vehicles on the apron.

Besides automating the docking process the video-based docking system has an additional advantage to airport administration. The sensor system recognizes the airplane leaving the parking position and can calculate the accurate parking time, which is then transmitted to the flight database of the airport.

5.6.3 Operation

A docking event can either be initiated manually by apron control using a control panel or be initiated fully automatic by receiving a command from the airport's flight database through a network. In order to activate the system, the expected aircraft type has to be selected. While approaching the parking position the pilots will follow the taxi-in line – a bright line painted on the apron ground – as close as possible. Using this information, the docking system is able to limit the area when the aircraft enters the view of the camera so that the image-processing system manages the successful initialization of a model-based tracking process. This initialization has to be done very fast with minimal latency, taking into account the limited amount of available computing resources. After initialization, the tracking process estimates the position and orientation of the aircraft in real-time.

The use of three-dimensional aircraft models guarantees the necessary stability of the tracking process and the demanded accuracy of the position estimation approach. By projecting the aircraft model from the three-dimensional scene into a two-dimensional image followed by an additional hidden-line removal step, the system calculates the expected aircraft outline for a predicted position of the plane (see Fig. 5.24). The predicted aircraft outline is compared with edge elements that are extracted from the HDR images by optimized image-processing algorithms. The tracking process iteratively improves the quality of the match between the expected outline and the edge elements detected in the sensor data and tries to find the optimal aircraft pose for the actual image within the available time slot. The system sends the calculated position to the pilot's display in order to guide the pilot to the stop position.

By using the three-dimensional aircraft models, the system benefits from the following advantages of model-based image sequence processing algorithms:

- The probability of a false initialization caused by passing vehicles is very low because they can be filtered out by the outline matching process.
- With the help of the aircraft's model information the system is able to assign all relevant image features and use them in the pose optimization process. This improves the reliability and the accuracy of the system in a

Fig. 5.24. Typical image of a docking sequence with the projected aircraft outline as overlay

significant manner especially during bad weather conditions with limited visibility when the edge detection algorithm can only find broken and separated fragments of the aircraft outline in the sensor information.
- The three-dimensional modeling of the scene allows extending the model knowledge by several characteristics of the scene, which under certain conditions can decrease the performance of the system. For instance, static obstacles in the scene, which are in the visual field of the camera and hide parts of the approaching aircraft, or shadow outlines caused by the aircraft itself, can be considered in the model and handled by the tracking process.
- The tracking process calculates all six degrees of freedom for the position and orientation of the aircraft. This allows for the camera to be placed anywhere without any major restrictions and enables the docking system to handle curved guidance lines.

One basic requirement to run a three-dimensional model-based tracking process is the accurate calibration of the camera. Position and orientation of the camera relative to the guidance line have to be measured precisely after setup.

5.6.4 Challenges to the Sensor System

The system uses a HDRC video camera to sense the approaching aircraft. The demands on the camera are enormous. The sensor has to deliver an image with sufficient contrast for the image-processing algorithms at all times and under all weather conditions. The camera's visual field is determined by the application: all types of aircrafts regardless of size being able to dock at the docking position need to be captured in their entirety by the camera while they are following a large section of the guidance line to their parking positions. On the other hand, in most cases the mounting point of the camera is predefined by the structure of the gate and does not allow optimizing the camera position. Depending on the compass direction of the guidance line, it is unavoidable that at some time of the day the sun will enter the visual field of the camera.

The situation for the sensor system becomes even more difficult at night when several vehicles in the vicinity will blind the camera with their headlights. The sensor must be able to handle sudden changes in illumination, for instance when the approaching aircraft switches off its landing headlamps during the docking process. In addition, the camera will be blinded by the reflections from different light sources on the wet apron during rainy nights.

On a sunny day differences in lighting within the scene, for instance between the reflective top surface of the aircraft body fully glared by the sun and the landing gear being completely in the dark shadow of the aircraft body and receiving only a weak amount of illumination, demand the camera to handle a very high-dynamic-range to facilitate the image-processing system to detect the complete outline of the aircraft.

5.6.5 HDR Camera with Improved Sensitivity

For imaging applications during uncontrolled and difficult illumination situations, we offer the HDRC camera GEVILUX CAM1 (see Fig. 5.25). By using a high-dynamic-range CMOS sensor with logarithmic response, this camera can handle a dynamic range of more then 140 dB within a single image. The large dynamic range of the sensor allows the gain of the amplifier in the analog path to stay fixed even under the broad spectrum of lighting situations on the apron during day and night as well as all weather conditions. Because there is no need to adjust any parameters within the camera depending on the illumination of the scene, the image acquisition process is absolutely stable during sudden changes of illumination.

The logarithmic response of the sensor is optimally adapted to the image-processing algorithms:

Fig. 5.25. HDR CMOS camera GEVILUX CAM1 with improved sensitivity

- Within areas with low illumination the sensor has a high response. This is why weak contrasts in dark areas are amplified and can be detected better by the edge detection algorithm.
- With increasing illumination the response becomes flatter. Well-illuminated image segments normally yield enough contrast so all edges can be detected with high probability.
- The response is continuous and monotonic. Therefore, additional difficulties at the image-processing stage are avoided.

The response differences of the sensor pixels typical for CMOS sensors, the so-called "fixed-pattern noise," are corrected within the camera by means of a sophisticated digital approach. The resolution of the A/D-conversion and the width of the data paths in the fixed-pattern noise correction unit have a width of 12 bit and give enough headroom to guarantee a 10 bit image data output without any restrictions from the correction unit.

A camera with sufficient sensitivity is needed for an operation with minimal illumination at night. The sensitivity of the CMOS sensor can be increased by lowering the operating temperature of the sensor. The camera GEVILUX CAM1 integrates a Peltier cooler to stabilize the sensor temperature. A control loop based on a high-resolution temperature sensor keeps the sensor temperature at a programmable value. A specifically designed camera package transfers the heat generated at the hot side of the Peltier cooler to a heat sink mounted on the camera package. The installed heat sink depends on the application and the selected sensor temperature.

When choosing the camera resolution for a real-time image sequence processing system, the following points have to be considered:

- In order to guarantee the stability of the tracking process, we have to maintain a certain minimal frame rate depending of the object speed in the image. If the visual tracking process is the input of a control loop, the stability criteria of the control loop normally defines the minimum frame rate.
- The image data rate from the sensor is limited by the bandwidth of the analog path.
- To transfer the digital image data from the camera to an image-processing unit, a communication channel with sufficient bandwidth is needed. The cost for such a high-speed data channel increases quite drastically when the image data have to be transmitted across long distances.
- The computing power at the image-processing unit is always limited. The sensor resolution has to be adapted to the available processing power in order to guarantee a minimum frame rate.

The maximum resolution of the camera GEVILUX CAM1 is 768×496 pixels with a data rate of 12 MHz. The maximum frame rate is 30 Hz at full resolution. The active image section is programmable and applications can choose a higher frame rate at a reduced resolution.

6

Lenses for HDR Imaging

Hans-Joerg Schoenherr

A measuring instrument is known to be as good as its weakest component!

In electronic imaging and image-processing systems so far this has been determined mainly by the well-known CCD imaging sensor with its significant "yardstick for all things": relatively small dynamic range in, linear gray tone conversion, and lightness-dependent mixture of colors.

High-dynamic-range logarithmical converting HDRC imaging sensors with eye-like characteristics do not have these disadvantages. However, they require high-dynamic-range lenses, if they are to be applied correctly in speedily fluctuating naturally occurring lighting situations with intense spot-like sources and back lighting. As a result, extreme lens contrasts (more than 100,000:1!) must be handled in real-time operation without any kind of controlled aperture or exposure.

Accordingly, the expectations for camera and image-processing systems with such high-capacity sensors are tremendous but cannot be fulfilled with merely new "HDR Ideal Optics". In order to fully utilize the new sensor possibilities and avoid having to work with single components close to physical thresholds this means that all hardware and software components have to be adjusted to each other in a way that all components are optimized according to today's technological means on an equally high level.

It is no longer advisable nor sufficient to push a single hardware and software component to its maximum limit individually.

The image-processing task itself should be included into such an optimizing process and especially the object as system component with its ancillary conditions needs to be thoroughly analyzed and incorporated during system development with the goal to create a "best compromise" regarding all components of the system as a whole.

Most lenses on the market (developed for CCD sensors capacity!) do not generally fulfill the HDRC sensor requirement in terms of dynamics, because of stray light as well as diversified reflections.

The demonstrated exemplary recordings stem from an extensive comparative test at KST with a digital EYELUX camera including a HDRC VGA1 sensor developed by KST and a camera with HDRC VGAx sensor developed by GEVITEC. Tests were conducted on a test object with an overall contrast range of 128,000:1 (17 levels of exposures).

This test object consists of six so-called "standard photographic objects" with a contrast of 32:1 (five levels of exposures) across the field of view (Figs. 6.1 and 6.2) in the form of matt gray-scale plates consisting of 12 equally graded fields (3–70% remission) against a neutrally gray background (15%) in an ideally matt black environment (0%). Centrically arrayed in the gray plates, there are spot-like sources of halogen back light beams inside the gray-scale plates which have a luminance that is more than 4,000-fold the luminance of an "object's white" with 100% remission of the medium gray-scale plate (12 levels of exposures).

The reflective light illuminance of the test object is carried out by means of a halogen flood light of equal spectral allocation (the illuminance on the medium gray tone plate being 200 lx, on the outer gray tone plate fading to 20 lx toward the edges of the image).

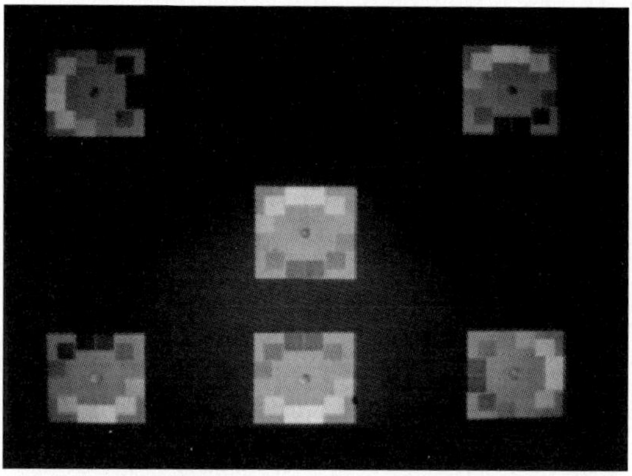

Fig. 6.1. Test object

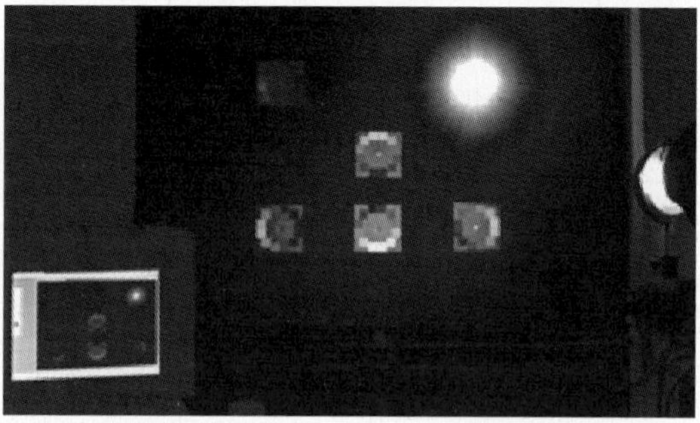

Fig. 6.2. Experimental set-up

The optical axis of the camera with the lens to be tested can be pivoted from 0° to 10° (see exemplary figures at 0°, 5°, or 10°, respectively) in order to simulate several different lighting situations and/or separate the reflex images from a direct image of the spot light source.

Typical development goals, ancillary conditions, and/or technically based settings generally aimed at silicon sensors and specifically aimed at CCD sensors were the unilateral focus during development of lenses in the past when correcting image errors like, for example:

- Highly reflective Si-sensor surfaces with a low coefficient for slanted beams
- Relatively low lens contrast of maximum 1:1,000 that can be recorded into one image by the CCD sensor
- Linear response curves of the gray tone conversion
- High intensity of light and equal illumination of the image field
- Sensor cover glass of the CCD in the optical path facing the image

These ancillary conditions lead to typical lens constructions with numerous free-standing lenses.

This trend was extended even further in recent years by the development of the highly effective antireflection coatings for glass–air surfaces (that are necessary to make such constructions possible!) as well as the demand for a more cost-effective automatic mass production of lenses.

When using HDRC sensors the remaining stray light of the annealed glass–air surfaces, lens edges, and frame surfaces whose intensity is generally below that of the registering threshold of the linear CCD sensors is intensified and mapped accordingly. Direct reflex images ("ghost images") of the shutter opening are created, more or less focused patterns of the shutter elements or the frame surfaces and a contrast reduction are created throughout the entire image field caused by diffuse stray light. (See images of current high-grade CCD top quality lenses for image processing and so-called mini lenses for platinum cameras designed by several manufactures with typically 5–8 free-standing lenses and/or lens elements (see Figs. 6.3–6.6), lenses with a fully opened shutter and partially dimmed by several F-stops.) The following

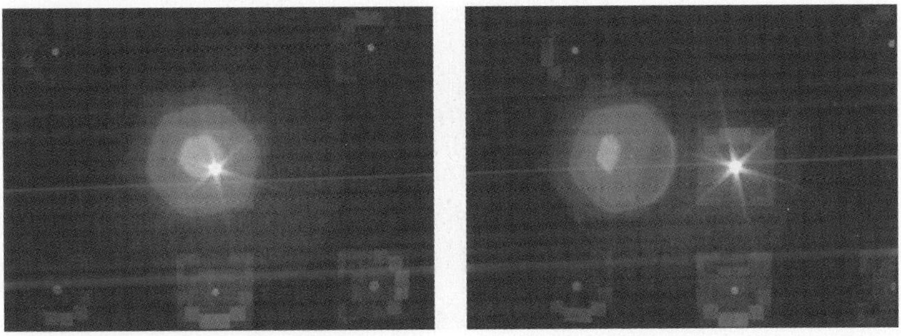

Fig. 6.3. Lens 1, 1:1.4, f 9 mm. *Left*: 0°. *Right*: 5°

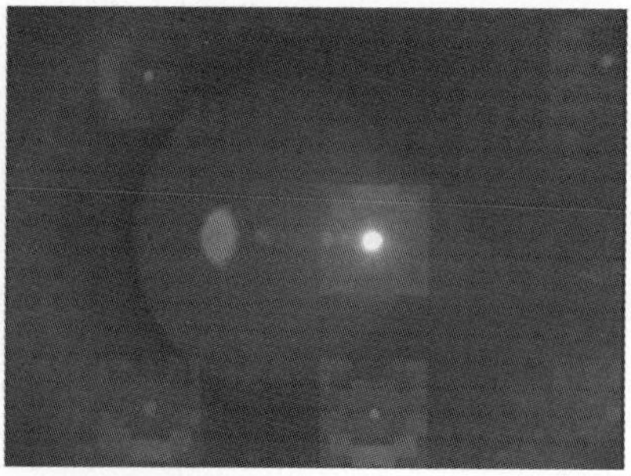

Fig. 6.4. Lens 2, 1:1.6, f 8 mm. *Left*: 0°. *Right*: 5°

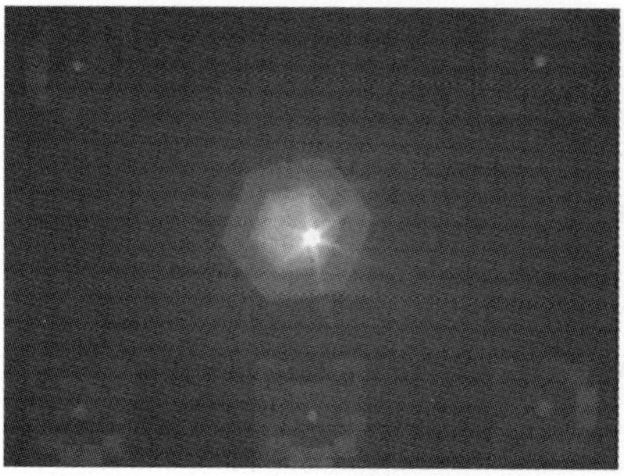

Fig. 6.5. Lens 3, 1:1.5, f 8.5 mm. *Left*: 0°. *Right*: 5°

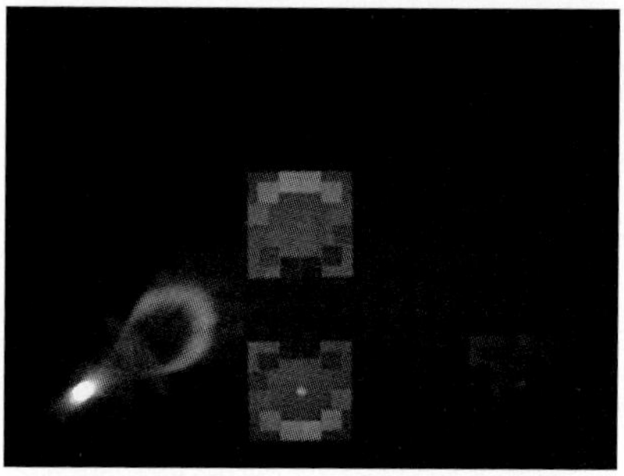

Fig. 6.6. Mini lens, f 6 mm

image processing is particularly distorted by the distinct "ghost images" that appear especially intense and focused for using aspheric lens surfaces when correcting image errors (see Figs. 6.7–6.9, mini lenses with asphere).

Basically, when developing lenses for high-dynamic-range sensors nothing "extraordinary" has to be developed that would not also serve other sensors' purposes as well. However, when choosing means for the design as well as when utilizing diversified latitudes for necessary corrections and selecting

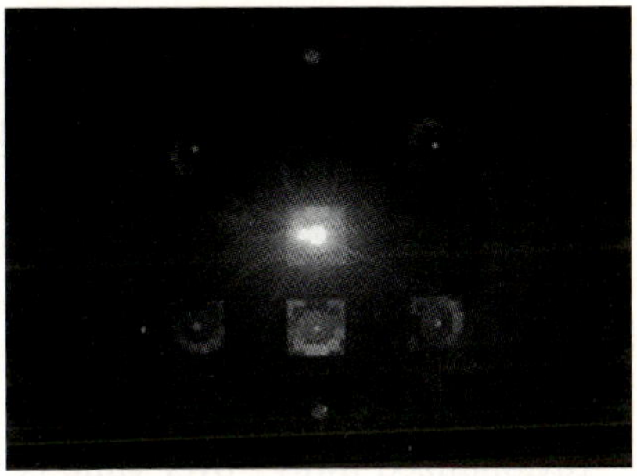

Fig. 6.7. Mini lens, f 4 mm, asphere, 0°

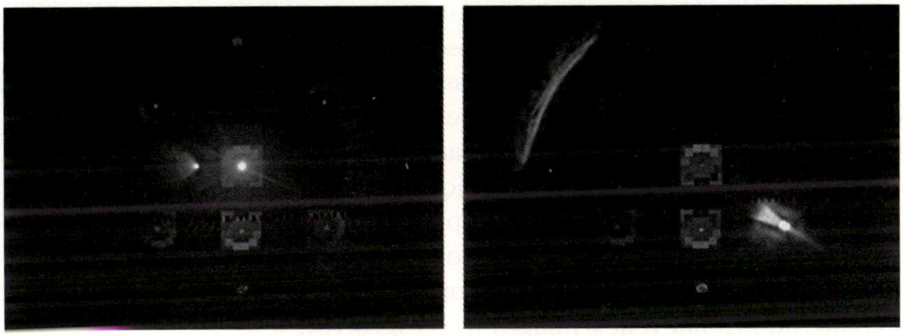

Fig. 6.8. Mini lens, f 4 mm, asphere, 5°

Fig. 6.9. Mini lens, f 10 mm, asphere, 5°

possible processing techniques and procedures different priorities will have to be applied!

Stray light should always be avoided at the source and not be dealt with a great deal of effort once it has appeared! This leaves the conclusion that, e.g., well-known virtues from early analog photography (despite far fewer technical possibilities!) like the self-understood maximum quality requirement that was realized by the optical industry regarding the minimizing of reflex images and stray light should once more become a priority, such as:

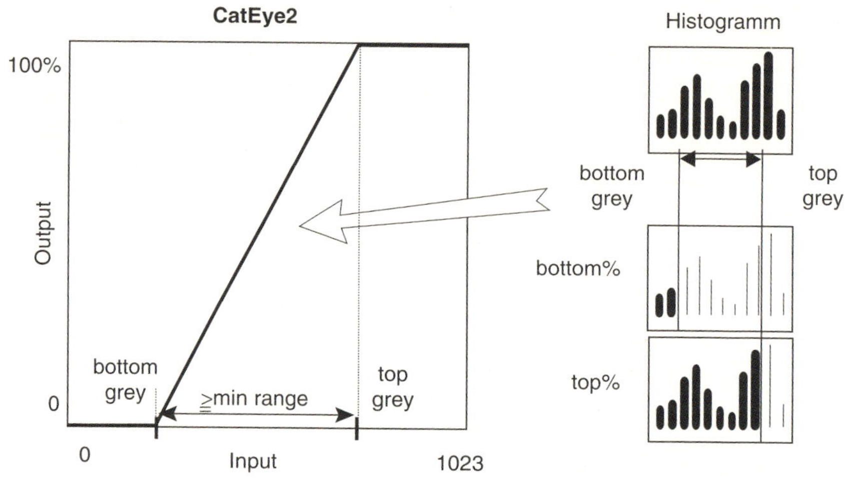

Fig. 7.6. Mode CatEye2

The detection of such blemishes can be very difficult because of the underlying ancillary conditions. Some of those are, for example, the strongly fluctuating degree of brightness on the surface to be tested and/or the area of shadow caused by the convolution of the surface. On the other hand, the impurities and/or surface discoloration contained in the lubricant (oil or lotion) during processing can create confusing appearances on the surface that is to be tested. These can then be detected as pseudodefects. In order to carry out a surface inspection successfully under these conditions, new algorithms are always required for defect detection. The physical basics for the creation of defects as well as adjustment mechanisms to deal with the fluctuation of brightness set the base for newly developed algorithms.

The hardware of an image-processing system, particularly the PC-based one, often cannot cope with calculation-intense real-time algorithms. In addition to that, the conventional CCD and CMOS cameras usually deliver a rather unusable image of metal surfaces. Such an image usually produces extremely under- or overexposed areas that can be traced back to brightness or an extremely corrugated form of the metal surface. Only a combination of high-performance evaluation electronics, a camera with a highly dynamic HDRC sensor and, adaptive evaluation algorithms can guarantee detecting all flaws on a surface with reflections and shades.

7.6 Evaluation Algorithms

Actually, the sought out defects are a "fluke of nature." They show certain characteristics that always occur with defects of the same type. The measurable values of these characteristics are, however, similar among two random defects but never identical. They can deviate quite significantly from the predetermined reference values. Therefore, these defects cannot really be described or ultimately recognized well with the assistance of formalistic methods, such as structural analysis [1,2] or edge profiling segmentation with predefined parameters [2, 3]. This is the reason for the necessity of always having new trouble-shooting methods in order to guarantee a thorough inspection of an inhomogeneous surface which can display either global or local fluctuations in brightness. For this, one needs to rely on the physical basics regarding the appearance of defects and the human visual behavior. Only an algorithm developed on this basis can be successful. Its implementation guarantees a definitive detection, recognition, and classification of surface defects. In order to develop such a procedure, the following basic question needs to be answered.

What is a Surface Defect?

As metallographic tests have shown repeatedly, a surface defect (crack, scratch, dent, etc.) is an object with characteristic edges [4]. These edges, however, are very similar to partial edges designed into the surface. Their similarity consists of an erroneous characteristic, such as the disproportional light between the surface and the background (erroneous depth) that is equal to the relevant characteristics across the partial edge. This is especially relevant for the width of the mapped edge that can be considered the junction width between the surface and the background (erroneous depth). This is independent of the luminous conditions across the edge. The characteristics of the edge to analyze itself as well as the parameters of the recording system, such as its resolution, however, determine the width of the recorded edge.

It is known [5] that a beam's intensity diversion displays a Gauss profile. Since a surface can be considered a beam source due to its reflexion the luminous diversion extracted from the surface across the edge to the background described as Gauss distribution [2]. This model has also proven to be the best one for the precise calculation of the edge position with subpixel accuracy [6]. This is why one can assume to have a Gauss profile (Fig. 7.7) when describing the gray value profile established across the edge. That means that this gray value profile has a normal diversion according to Gauss and, that is why, is an exponential function and, therefore, is characterized by certain specifically proportional points.

Accordingly, the following *adaptivity parameters* can be derived from the gray value profile:

- σ: the standard deviation representing the semiedge width of the detected edge
- ξ_1: the inflection point coefficient that determines the brightness I_{turn} on the edge
- ξ_2: the turning point coefficient that determines the background brightness I_{bgrd}

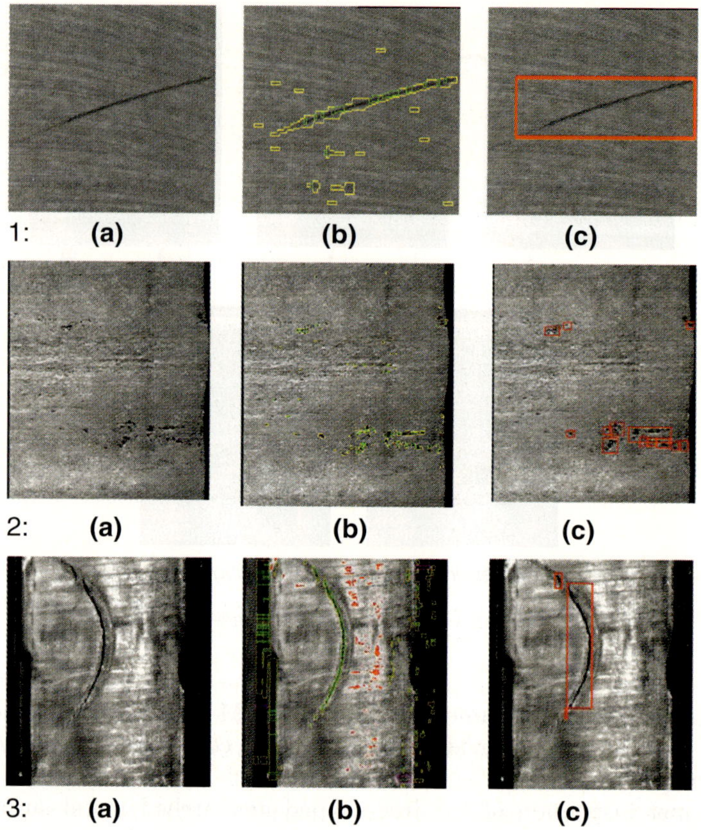

Fig. 7.9. Detection capacity of the newly developed algorithm on several metal surfaces with different: (1) massive ground surface with a scratch; (2) massive raw surface with several cages; (3) unwashed surface of a preformed part with a crack; (**a**) original image; (**b**) processing image with all candidates (*yellow marking*) and 100% defect positions (*green marking*); (**c**) resulting image with defect (*red frame marking*)

This means that an explicit and safe detection of all defects on different metal surfaces are guaranteed with the aid of an adaptive edge-based algorithm (Fig. 7.12).

7.7 Robot Controlled Image-Processing System for Fully Automated Surface Inspection

The combination of high-performance evaluation electronics, a camera with a highly dynamic HDRC image sensor and an adaptive evaluation algorithm produces a product for the inspection of free shaped metal surfaces – the MetalDefectDetector. The MetalDefectDetector is implemented with an intelligent

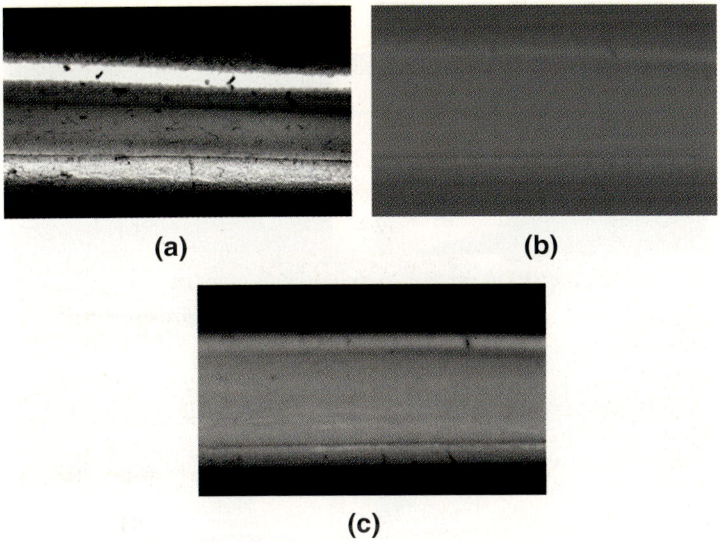

Fig. 7.10. Deformed metal surface with a constriction: (**a**) recording with a CCD camera (*original image*); (**b**) recording with a ICAM-HD1(*original image*); (**c**) recording with a ICAM-HD1(*enhanced image*)

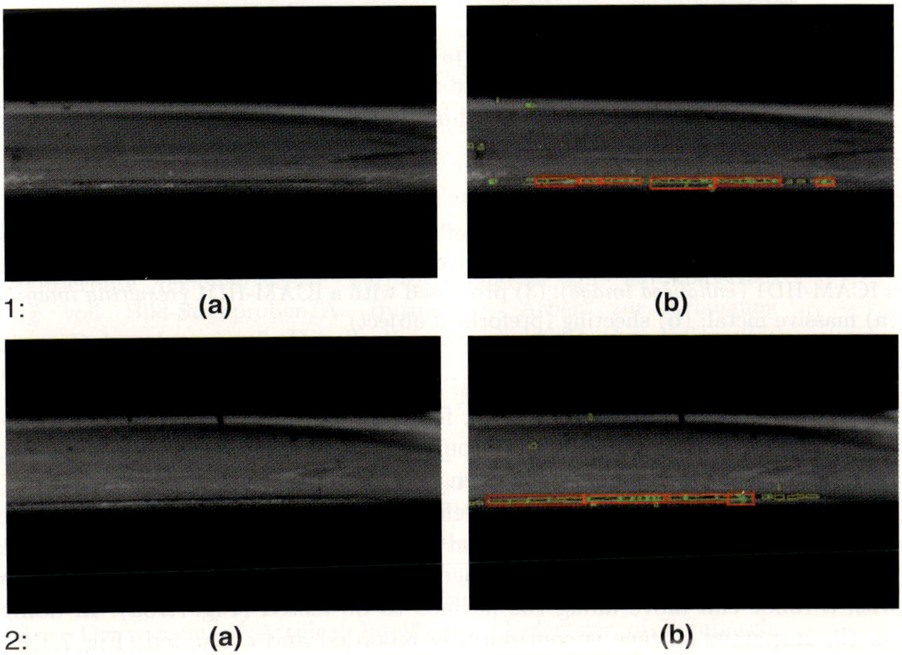

Fig. 7.11. Defect detection capacity of the adaptive edge-based algorithm on on a deformed metal surface (recording with a ICAM-HD1): (1) closed constriction; (2) closed constriction (*tear*); (**a**) enhanced image; (**b**) resulting image (with all marks)

- A high sensitivity is required especially in the case of image presentation to the driver
- A wide dynamic range (>120 dB) is required to cover the high brightness range
- A nonlinear intensity response is required to avoid blinding of the imager by the headlights of oncoming vehicles

To achieve a naturalistic image on the graphic display, corrections and additional signal processing are required [8]:

- CMOS imagers have a fixed pattern noise (FPN), needing an additional memory and a calibration procedure
- After grabbing the image additional noise reduction has to be made
- Edge enhancement is necessary
- A nonlinear intensity compression is required

8.8.1 Lane Detection and Lane Departure Warning

Lane departure warning (LDW) systems can be operated with a mono or with a stereo camera. As shown in Fig. 8.11, the camera searches for lines in the front of the vehicle. The upper image shows the camera image with search lines (see detail in the second image from above) and the crosses in the upper image show the lane course, which is calculated by the image processing computer.

To detect a line, the luminance signal within the search line is analyzed and, using a high pass-filtering algorithm, the edges of the line are detected.

From these signals a warning can be derived if the driver is crossing the lane mark without having set his turning lights. This warning may be acoustical or haptical, e.g., by giving a momentum on the steering wheel in the opposite direction [9].

8.8.2 Traffic Sign Recognition

A traffic sign recognition system consists of the above-described video system but with a mono camera. The image of the camera is shown in Fig. 8.12. While driving, the camera searches permanently for round shaped objects, which could be a traffic sign. When such an object is detected, the object is tracked until the resolution of the camera chip is sufficient for reading the sign. Reading in this case means a comparison between the recognized sign and sign patterns which have previously been learned by the system.

The recognized traffic signs can then be displayed on a graphic display being located in the instrument cluster or in the center console. It is easy to add an acoustic warning when the driven speed is higher than the allowed speed [9].

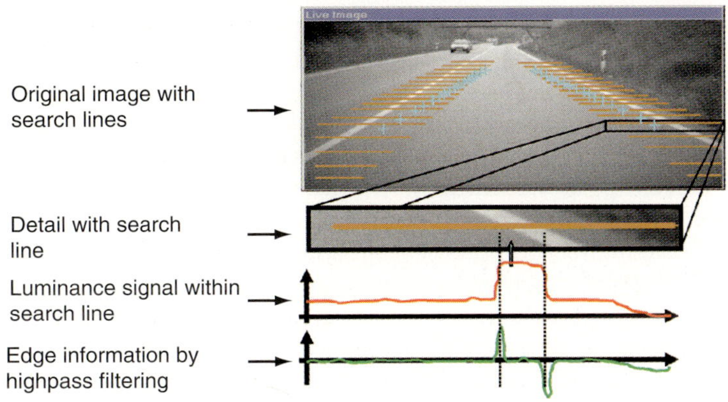

Fig. 8.11. Principle of lane detection

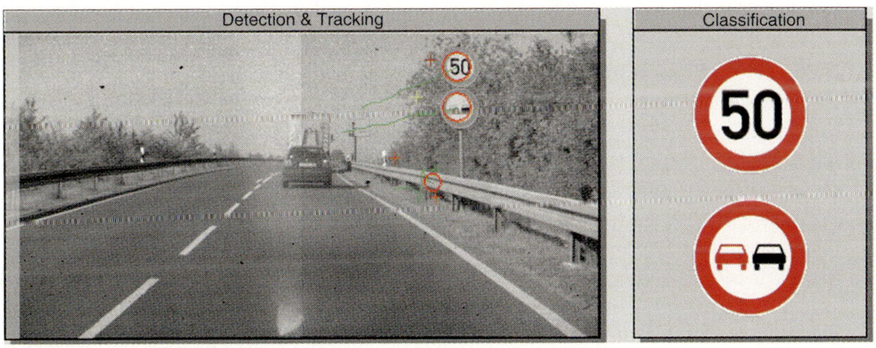

Fig. 8.12. Traffic sign recognition: camera image with tracking lines and recognized signs

8.9 Conclusion

Driver assistance functions have the potential to significantly contribute in reducing accidents and fatalities while driving at day and night. As basic technology with most advantages for future functions seen from the evolution side of night vision systems, the active near-infrared technology combined with a high-resolution CMOS-sensor is expected.

Current first generation night vision systems presenting live images in the primary field of the driver's view will be substituted by systems based on object detection. These second generation systems will provide warnings instead of images to the driver, e.g., for lane departure warning, lane guidance, or obstacle collision warning, which will require new ideas for human machine interfaces.

From this stage on, the introduced functions will make further use of data fusion with other sensors or systems, like navigation maps and range sensors based on RADAR or LIDAR technology, to increase the reliability in detecting objects. Further on, object detection will be the basis for controlling actuators for lane keeping functions and collision mitigation and collision avoidance functions.

The highest demand regarding performance and reliability is put on active safety systems. They range from a simple parking stop, which automatically brakes a vehicle before reaching an obstacle, to computer-supported control of complex driving maneuvers to avoid collisions. For example, the automatic emergency braking feature intervenes if a crash is unavoidable. In its highest levels of refinement, active systems intervene in steering, braking, and engine management to avoid colliding with an obstacle. Here, the vision goes to the collision avoiding vehicle, making computer assisted driving maneuvers for crash avoidance [5]. For all these future functions, video is a key technology.

References

1. Enke, K.: "Possibilities for Improving Safety Within the Driver Vehicle Environment Loop", 7th Intl. Technical Conference on Experimental Safety Vehicle, Paris (1979)
2. Knoll, P.M.: "Mehr Komfort und mehr Sicherheit durch prädiktive Fahrerassistenzsysteme", Conf. Proc. Vision Automobil, Handelsblatt, München (2005)
3. Seger, U.; Knoll, P.M.; Stiller, C.: Sensor Vision and Collision Warning Systems, Convergence Conference, Detroit (2000)
4. Knoll, P.M.: "Predictive Safety Systems" Mstnews, Mitteilungsblatt der DLR zum Förderprogramm Mikrosystemtechnik (2004)
5. Schäfer, B.-J.; Knoll, P.M.: "Prädiktive Fahrerassistenzsysteme – vom Komfortsystem zur aktiven Unfallvermeidung", Proc. VDI-Conference, Wolfsburg (2004)
6. Knoll, P.M.: Fahrerassistenzsysteme – Realer Kundennutzen oder Ersatz für den Menschen? VDI, Deutscher Ingenieurtag, Münster, Germany (2003)
7. Anonimous statistics of accident data of the "Statistisches Bundesamt" (German Federal Statistical Institution), Wiesbaden, Germany (2002)
8. Bischoff, S.; Haug, K.: "Automotive NIR-Night-Vision-Systems: Challenges for Series Introduction", Proc. Vision 2004, Rouen (2004)
9. Knoll, P.M.: "Handbuch Kraftfahrzeugtechnik", chapter "Fahrerassistenzsysteme", Vieweg-Verlag (2005)

9

Miniature HDRC Cameras for Endoscopy

Christine Harendt and Klaus-Martin Irion

Visualization of the status of organs health is one major task in medical diagnosis and therapy. Endoscopy and Minimal Invasive Surgery (MIS) are techniques for this purpose in medical applications. Recent developments in microelectronics allow the fabrication of advanced and highly integrated image sensors and improved solutions for miniaturization and wireless data transmission [1].

In endoscopy important aspects are the demand for smaller devices and the need to integrate high quality, but low cost visualization techniques. Moreover, the problems and cost of sterilization raised the wish to fabricate disposable endoscope heads.

While the majority of endoscopes uses optical lens systems (rigid endoscopes) or fiber bundles (flexible endoscopes) to transmit the images to a camera, video-endoscopes have the camera directly at the tip. Thus, such an endoscope head is a micro system with image sensor chip, optics, illumination and electrical wiring. Image data is transmitted via cables, which also provide the power for the system.

A major objective in a recently finished research project of the European IST (Information Society Technology) program was the development of a small image sensor for an endoscope with a very small diameter. In the project IVP (Intracorporeal Videoprobe) a consortium with partners from four European member states cooperated in the development of videoprobes [2]. One prototype, the so-called wired probe IVP1, has a diameter of the endoscope head of 3.5 mm (Fig. 9.1).

The HDR color image sensor could be realized with outer dimensions of $1.7 \times 1.3\,\mathrm{mm}^2$and a resolution of 36,000 pixels.

A major issue in the chip development was the size reduction of both the image area and the overall chip. Thus, additional to the general objectives (sensitivity, dynamic range and low dark current) three issues had to be addressed:

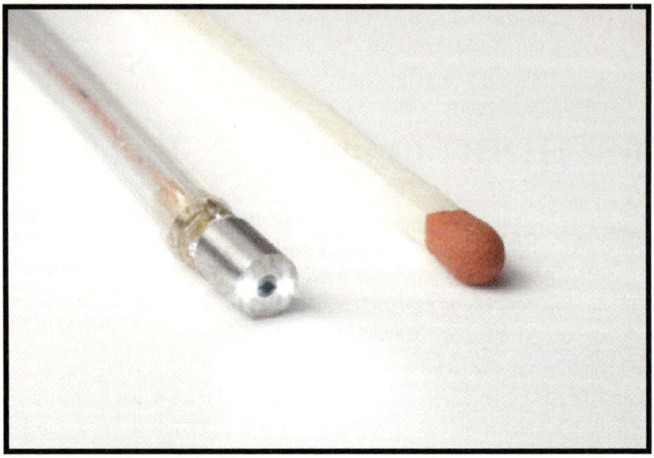

Fig. 9.1. Videoprobe with optics and illumination

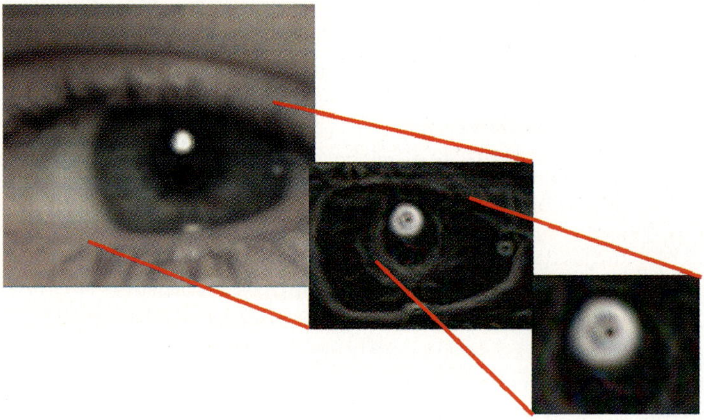

Fig. 9.2. Color image obtained with an IVP1 image sensor. The enlarged images show details revealed after edge detection

- The minimization of the pixel size to allow acceptable dimensions for this sensor.
- The reduction of the number of pad connections because bonding pads with their area and the required sizes for pad circuitry cover a significant area of the chip. Additionally, each bond also requires size on the package.
- The testability of the device to allow complete evaluation and verification of the first devices.

The resulting sensor with a logarithmic response has a pixel pitch of 4.6 µm. The sensor shows a measured dynamic range >100 dB and has a minimal detectable illumination of 0.03 lx.

Figure 9.2 demonstrates the performance of the color device on an image of a human eye. Although the bright reflection in the center seems to be saturated, more details are visible in the enlarged images, which have been processed with an edge-detection software. The details are the shape of the spot light used for the illumination during the shot.

The assembled device has only four connections, but the chip is originally equipped with 43 pads, which are used for the complete digital test. The majority of these pads are cut-off before the final mounting of the chip.

For the fabrication of the IVP1 prototype the chip is mounted on a circular (Ø 3 mm) ceramics, where it is connected to the cables. Two openings in the circuit board are used to fix the optical fibers for the illumination: a metal cap with the optics completes the distal end of the endoscope. The analog image data are transferred via the cables to a PC, where the image data is processed and subsequently displayed.

References

1. K.M. Irion "Endoscopes – Potential for Microsystem Technology" MST News, Special Issue "Microsystems Made in Germany" 2005, p 26
2. A. Arena, M. Boulougoura, H.S. Chowdrey, P. Dario, C. Harendt, K.M Irion, V. Kodogiannis, B. Lenaerts, A. Menciassi, R. Puers, C. Scherjon, D. Turgis, "Intracorporeal Videoprobe (IVP) " in Medical and Care Compunetics 2, Volume 114, edited by L. Bos, S. Laxminarayan and A. Marsh, May 2005, IOS Press, pp. 167–174

there is a NAND an inverter gate for cell selection and some switches needed to connect the output of the amplifier to the test-circuit and to make the inputs accessible.

A pulsed power supply is connected to the amplifier. With each pulse, the amplifier cell emits charge between 0.5 and 10 nC to the high capacitive load connected to the tissue. The maximum height of the output pulses is set to 2 V because a higher voltage can cause electrolysis effects. For this reason, the design of the amplifier considers 2 V as the maximum output voltage of the amplifier. The electrode capacitance restricts the charge margin.

Discharging the output node to ground reference voltage during the time of inactivity secures charge balance at full pulse cycle. This is done by a pull-down circuit that forces the amplifier outputs to ground during inactivity and thus discharges the electrodes. Additionally, this implements charge balance regarding the stimulation impulses.

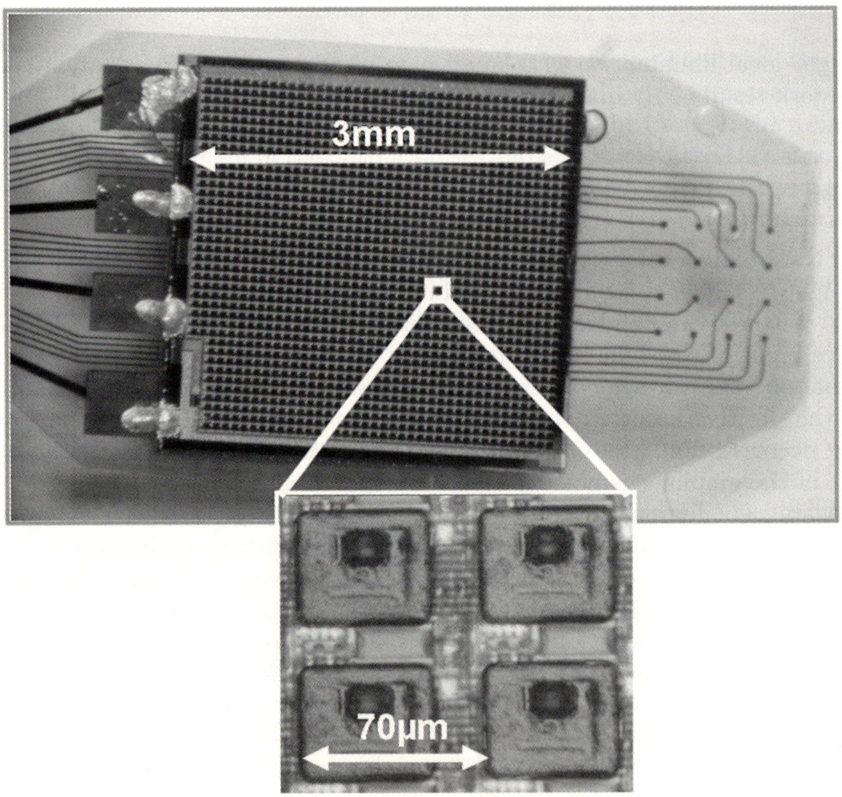

Fig. 10.3. Capsulated retinal implant. The higher magnification shows the output contacts to the tissue (photo: Retina Implant AG)

The pull-down signal is generated by an NMOS inverter. The inverter is connected to a different supply voltage to keep on working even if the pulsed supply is switched off. The dimensions of the cells are $72 \times 72\,\mu m^2$.

10.5 The Retinal Implant

The complete implant consists of 38×38 cells. The space of the missing cells is needed to place the reference source, the pull-down inverter and the bonding pads to connect the final implant. The dimension of the implant chip is $3 \times 3.1 mm^2$ (Fig. 10.3).

During normal operation, the implant uses a pulsed supply at about 20 Hz. The active time per period is about 500 µs. The pull-down signal is generated by means of an additional DC supply voltage.

Acknowledgement

The project, in which this implant was developed, was funded in part by BMBF (German Federal Ministry of Education and Research). The co-operation with NMI (Naturwissenschaftlich Medizinisches Institut, Reutlingen) with the Augenklinik, University Tuebingen and with Retina Implant AG, Reutlingen, is gratefully acknowledged.

References

1. E. Zrenner, Will retinal implants restore vision, Science, 295, 1022–1025, Feb. 2002
2. W.H. Dobelle, Artifcial vision for the blind by connecting a television camera to the visual cortex, ASAIO-J, 46, Jan–Feb 2000, 3–9
3. C. Veraart et al., Brain Res. 813, 181, 1998
4. R. Eckmiller, Learning retina implants with epi-retinal contacts, Ophthalmic Res., 29, 281–289, 1997; A.Y. Chow, M.T. Pardue, V.Y. Chow, G.A. Peyman, C. Liang, J.I. Perlman, N.S. Peachey, Implantation of silicon chip microphotodiode arrays into cat sub-retinal space, IEEE J. Solid State Circuits, 35(10), 2000
5. M. Schwarz, R. Hauschild, B.J. Hosticka, J. Huppertz, et al., Single-chip CMOS image sensors for a retina implant system, IEEE Trans. Circuits Systems II, 46(7), 1999
6. R.J. Greenberg, T.J. Velte, M.S. Humayun, G.N. Scarlatis, E. de Juan, A computational model of electrical stimulation of retinal ganglion cell, IEEE J. Solid State Circuits, 35(10), 2000
7. A.Y. Chow, Electrical stimulation of the rabbit retina with subretinal electrodes and high density microphotodiode array implants, Invest. Ophthalmol Vis. Sci. 34(suppl), 835, 1993

8. M. Stelzle, A. Stett, B. Brunner. M. Graf, W. Nisch, Electrical properties of micro-photodiode arrays for use as artificial retina implant, Biomed. Microdevices, 3(2), 133–142, 2001
9. B. Schlosshauer, A. Hoff, E. Guenther, E. Zrenner, Towards a retina prosthesis model: Neurons on microphotodiode arrays in vitro, Biomed. Microdevices, 2(1), 61–72, 1999
10. F. Geckler, H. Schwahn, A. Stett, K. Kohler, E. Zrenner, Sub-retinal microphotodiodes to replace photoreceptor function. A review of the current state, Vision, sensations et environment, Irvinn, Paris, 2001, pp. 77–95
11. A. Dollberg, H.G. Graf, B. Höfflinger, W. Nisch, J.D. Schulze Spuentrup, K. Schumacher, E. Zrenner, A fully testable retinal implant, Proceedings of IASTED Conference on Biomedical Engineering, Salzburg 2003, pp. 255–259

11

HDR Tone Mapping

Grzegorz Krawczyk, Karol Myszkowski, and Daniel Brosch

High dynamic range (HDR) imaging is a very attractive way of capturing real world appearance, since it permits the preservation of complete information on luminance (radiance) values in the scene for each pixel. This, however, requires the development of a complete pipeline for HDR image and video processing from acquisition, through compression and quality evaluation, to display. While the prospects for proliferation of HDR capturing (refer to Chap. 2 and 5) and multimedia (refer to Chap. 12) technologies are good, an important problem to solve regards how to reproduce the scene appearance using media with very limited dynamic range such as hard-copy prints, CRT/LCD displays, and projectors. While this may change in the future due to successful efforts in building HDR displays [1] (refer also to Chap. 14) at present such devices are relatively rare and expensive. To solve the problem of transforming high dynamic range images to the limited range of most existing devices, the so-called tone mapping operators are being developed. Essentially, tone mapping addresses the problem of strong contrast reduction from scene values to displayable ranges while preserving the image details, which are important to appreciate the scene content.

The goals of tone mapping can be differently stated depending on the particular application. In some cases producing just "nice-looking" images is the main goal, while other applications might emphasize reproducing as many image details as possible, or might maximize the image contrast. The goal in realistic rendering applications might be to obtain a perceptual match between a real scene and a displayed image even though the display device is not able to reproduce the full range of luminance values. However, achieving such a match is feasible because the human visual system (HVS) has limited capacity to detect differences in absolute luminance levels, and concentrates more on aspects of spatial patterns when comparing two images [2].

It should be noted that the choice of an appropriate tone mapping algorithm and its parameters may depend not only on a particular application, but also on the type of display device (projector, plasma, CRT, LCD) and its characteristics, such as reproduced maximum contrast, luminance range, and gamma correction. Also, the level of surround luminance, which decides upon the HVS adaptation state and effective contrast on the display screen, is important in this context. This means that the visual quality of displayed content can be seriously degraded when already tone mapped images and video are stored or transmitted without any prior knowledge of their actual display conditions.

The design of a typical tone mapping operator is usually guided by few rules. It must provide consistent results despite the vast diversity of natural scenes and the possible radiance value inaccuracy found in HDR photographs. Additionally, it should be adaptable and extendible to address the current capabilities of displaying methods and their future evolution. Tone mapping must capture the physical appearance of the scene, while avoiding the introduction of artifacts such as contrast reversal or black halos. The overall brightness of the output image must be faithful to the context. It must be "user-friendly" i.e., automatic in most cases, with a few intuitive parameters which provide possibility for adjustments. It must be fast for interactive and realtime applications while avoiding any trade-off between speed and quality.

In this chapter we briefly overview existing tone mapping operators (Sect. 11.1). Since a majority of those operators have been designed for static images with the assumption of ideal (noise-less) sensors used for their capturing, in Sect. 11.2 we discuss specific requirements imposed on tone mapping for real-world HDR video. We present in more detail a complete solution suitable for real-time HDR video tone mapping in Sect. 11.3. In Sect. 11.4 we discuss perceptual effects such as glare, rod-vision that cannot be evoked by typical displays because of their limited dynamic range, but can be easily simulated in real-time atop of a video stream based on the knowledge of HDR luminance and contrast levels. In Sect. 11.5 we discuss a high quality tone mapping operator based on bilateral filtering, which due to high computational cost is suitable only for off-line HDR video processing. Finally, in Sect. 11.6 we summarize major requirements imposed on tone mapping operators in HDR video applications.

11.1 Taxonomy

In general, tone mapping can be classified as *time dependent* for handling animated sequences and *time independent* designed specifically for static images. In the former case temporal adaptation conditions of the HVS such as dark or bright adaptation are taken into account.

Another classification takes into account whether the same mapping function is used for all pixels (*spatially uniform* or *global* operators) or if the mapping function variates depending on a given pixel neighborhood which is often achieved through the modeling of spatial adaptation (*spatially varying* or *local* operators). In the former case a compressive, monotonic tone reproduction curve (TRC) is constructed which for a given HDR pixel luminance level always assigns the same output pixel intensity. For spatially varying operators the output pixel intensity depends also on the scene surround and can vary quite significantly even for pixels with the same luminance level.

Yet another classification takes into account whether a given tone mapping algorithm tries to model some characteristics of the HVS or perhaps is a pure image processing technique which essentially transforms one image into another. In the former case through mimicking the HVS processing it is expected to achieve *predictive* tone mapping whose the main goal is to obtain a good match between the real world scene and displayed scene percepts. The objective of image processing approach is often to obtain the results which are *preferred* by the observer, even if they depart from realism and are not plausible in terms of the real world appearance.

11.1.1 Spatially Invariant Operators

In this section we overview global tone mapping operators which are inspired by psychophysical models of brightness and contrast perception (Brightness and Contrast Perception) as well as psychophysiological models of retinal photoreceptors (Photoreceptor Model of Contrast Compression). Then we discuss more engineering oriented solutions for contrast compression using logarithmic functions (Logarithmic Contrast Compression) and image histogram adjustment (Histogram Adjustment Methods).

Brightness and Contrast Perception

The term brightness B describes the response of the human visual system (HVS) to stimulus luminance Y. This response has the form of compressive nonlinearity which can be approximated by a logarithmic function (Weber-Fechner law) $B = k_1 \ln(Y/\bar{Y})$, where $\bar{Y}$ denotes the luminance of the background and k_1 is a constant factor. This relation has been derived in psychophysical threshold experiments through examining just noticeable differences (JND) ΔY for various $\bar{Y}$. Slightly different relations between B and Y have been obtained depending on such factors as the stimulus size, the luminance of adaptation $\bar{Y}$, and temporal presentation. For example, suprathreshold experiments resulted in an observation that equal ratios of luminance lead to equal ratios of brightness and the HVS response should be rather modeled by a power function (Stevens law) $B = k_2 Y^n$, where n falls in the range of 0.3–1.0. In practice, the Weber law is a good approximation for bright illumination conditions ($\bar{Y} > 1{,}000\,\mathrm{cd{\cdot}m^{-2}}$) and for lower adaptation luminance levels ($0.1 < \bar{Y} < 1{,}000\,\mathrm{cd{\cdot}m^{-2}}$) power functions fit better to the experimental data (refer also to Sect. 12.6).

Tumblin and Rushmeier [3] have developed a perception-inspired tone mapping whose the main objective is to preserve a constant relationship between the brightness of a scene perceived on a display and its real counterpart, for any lighting condition. They use Stevens' power-law relationship to transform the real world luminance into the real world brightness. Then based on the brightness matching assumption they find the corresponding display luminance values using inverse display observer and inverse display device transformations.

Since the Tumblin and Rushmeier operator is derived from suprathreshold psychophysical data it predicts well the brightness magnitude, but may be imprecise in predicting whether some low contrast details in the real world scene can be perceived by the human observer. This task is relatively easy for tone mapping operators [4,5] based on the psychophysical measurements of the threshold vs. intensity (TVI) function which for each luminance adaptation level $\bar{Y}$ specifies just noticeable luminance threshold ΔY. It is assumed that the eye is adapted to unique luminance levels $\bar{Y}_{\mathrm{w}}$ and $\bar{Y}_{\mathrm{d}}$ corresponding to the real world and to the display observation conditions ($\bar{Y}$ is usually computed as the average of logarithm values of each pixel luminance in the image). Then the corresponding just noticeable threshold values ΔY_{w} and ΔY_{d} are found for $\bar{Y}_{\mathrm{w}}$ and $\bar{Y}_{\mathrm{d}}$ based on the TVI function. The tone mapping between the real world luminance Y_{w} and display luminance Y_{d} is then a simple scaling function:

$$Y_{\mathrm{d}} = \frac{\Delta Y_{\mathrm{d}}}{\Delta Y_{\mathrm{w}}} Y_{\mathrm{w}}. \tag{11.1}$$

Due to the linear nature of this operator it is not suitable for images of very high dynamic range and it poorly predicts contrast magnitudes well over the perceivability threshold.

Photoreceptor Model of Contrast Compression

Neurological and psychophysiological studies of HVS demonstrate that nonlinear luminance compression can mostly be attributed to the retinal signal processing. Direct measurements of cellular response functions for cones, rods, bipolar, and ganglion cells show that they belong to the family of sigmoid functions

$$R(Y) = R_{\mathrm{max}} \frac{Y^\alpha}{Y^\alpha + \sigma^\alpha}, \tag{11.2}$$

where R is the neural response ($0 < R < R_{\mathrm{max}}$), Y is the luminance value, σ is the semi-saturation constant which corresponds to $\bar{Y}$ for which $R(\bar{Y}) = 0.5 \cdot R_{\mathrm{max}}$. The response sensitivity is controlled by α which is found to lie between 0.2 and 0.9. The response function $R(Y)$ is close to linear for luminance close to the adaptation level $\bar{Y}$ and features strong compression of luminance values significantly departing from $\bar{Y}$ (both for extremely dark and bright pixels).

The photoreceptor model in the form of (11.2) with $\alpha = 0.73$ is the foundation of time-dependent tone mapping operator developed by Pattanaik et al. [6]. This operator deals with the changes of threshold visibility, color appearance, visual acuity, and sensitivity over time. The operator can be decomposed into two models: the visual adaptation model and the visual appearance model. The signals that simulate the adaptation measured in the retina are used for adaptation in each pixel of an image. To reproduce visual appearance, it is assumed that a viewer determines reference white and reference black colors. Then, the visual appearance model recalculates the visual appearance with respect to those reference points. By assembling the visual adaptation and appearance models, the scene appearance is reproduced with changes to visual adaptation depending on time. This method is also useful to predict the visibility and appearance of scene features because it deals with reference white and black points.

Reinhard and Devlin [7] used in their tone mapping operator a version of photoreceptor model as proposed by Hoods and colleagues. An important contribution of their work is the control of color saturation in the spirit of a von Kries model.

A sigmoid function is also used in the photographic tone reproduction which belongs to the category of spatially variant tone mapping operators (refer to Photographic Tone Reproduction and Sect. 11.3.1).

Logarithmic Contrast Compression

Stockham [8] recommended a logarithmic relation in his early tone mapping solution

$$R(Y_{\mathrm{w}}) = \frac{\log(Y_{\mathrm{w}}+1)}{\log(Y_{\max}+1)}, \tag{11.3}$$

where for each pixel, the output intensity $R(Y_{\mathrm{w}})$ is derived from the ratio of world luminance Y_{w} and maximum luminance in the scene $Y_{\max}$. This mapping ensures that whatever the dynamic range of the scene is, the maximum value is remapped to one (white) and other luminance values are smoothly changing.

While the Stockham approach leads to pleasant images, Drago et al. [9] found that the luminance compression is excessive and the feeling of high contrast content is lost. Drago et al. introduced a method which is called adaptive logarithmic mapping. This method addresses the need for a fast algorithm suitable for interactive applications which automatically produces realistically looking images for a wide variation of scenes exhibiting high dynamic range of luminance. To preserve details while providing high contrast compression, a family of logarithmic functions with increasing compressive power are used

$$\log_{\mathrm{base}}(Y) = \frac{\log(Y)}{\log(\mathrm{base})} \tag{11.4}$$

with $\log_{\mathrm{base}}$ ranging from $\log_2$ to $\log_{10}$. The $\log_{10}$ is applied for the brightest image pixel and for remaining pixels the logarithm base is smoothly interpolated between values 2 and 10 as a function of their luminance:

$$R(\hat{Y}_{\mathrm{w}}) = \frac{\log(\hat{Y}_{\mathrm{w}}+1)}{\log\left(2+8\cdot\left(\frac{\hat{Y}_{\mathrm{w}}}{\hat{Y}_{\mathrm{w\,max}}}\right)^{\frac{\log(b)}{\log(0.5)}}\right)}, \tag{11.5}$$

where each pixel luminance $\hat{Y}_{\mathrm{w}}$ and the maximum luminance of the scene $\hat{Y}_{\mathrm{w\,max}}$ are normalized (divided) by the world adaptation luminance $\bar{Y}_{\mathrm{w}}$. Perlin bias power function [10] in the denominator of (11.5) is used for interpolation between the logarithm bases to provide better steepness control of the resulting tone mapping curve. The bias parameter b is essential to adjust compression of high values and visibility of details in dark areas. Values for b between 0.7 and 0.9 are most useful to generate perceptually good images. Psychophysical experiments with human subjects suggest that the Drago et al. operator leads to images which are highly ranked in terms of natural appearance and subject preferences [11].

Histogram Adjustment Methods

Ward Larson et al. [12] proposed the histogram adjustment method which allocates most of the displayable dynamic range to the luminance ranges which are represented by many pixels in the scene. This results in strong contrast compression for pixels belonging to sparsely populated regions in the image histogram, which helps to overcome the problem of dynamic range shortage. On the other hand, for HDR images with more uniform distribution of pixels across the full luminance range the compression would be weak. The method leads to a monotonic tone reconstruction curve which is applied globally to all pixels in the image. The slope of the curve is constrained by considering the human contrast sensitivity to guarantee that displayed image does not exhibit more contrast than it could be perceived in the real scene. The method incorporates also other characteristics of the HVS such as color sensitivity, visual acuity, and glare.

Comparison of Global Mapping Curves

In Fig. 11.1 we show for comparison several tone mapping curves adjusted to a sample HDR image. The histogram of the image is plotted in the background for reference. Clearly, a naive approach to linearly scale the luminance values is not able to map the input dynamic range to displayable values and the details in high and low luminance ranges are lost. The full range of input luminance values can be successfully mapped using the adaptive logarithmic mapping [9] (refer to "Logarithmic Contrast Compression"). Equally good performance can be achieved with the photoreceptor model [7], though the image will be considerable brighter as the mapping curve maps low luminance values to higher display values (refer to "Photoreceptor Model of Contrast Compression"). The curve of the histogram adjustment method [12] not only covers the whole input range but also fine tunes its shape to assign wider displayable range to highly populated luminance ranges what in the end effect produces a well optimized final result (refer to "Histogram Adjustment Methods").

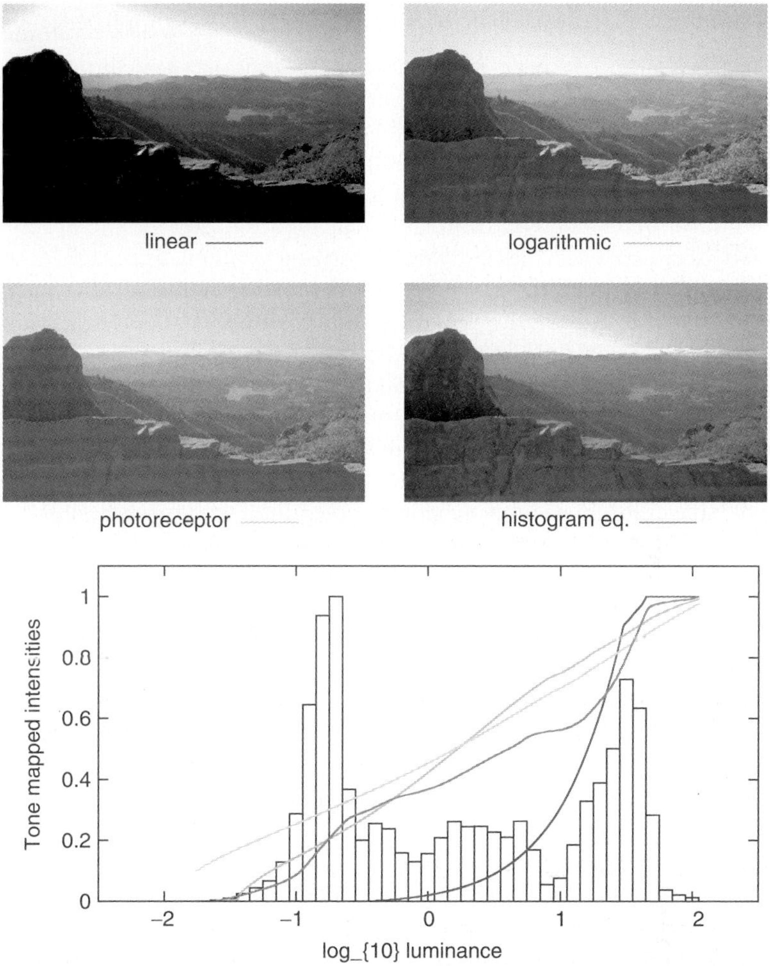

Fig. 11.1. Various mapping curves of global tone mapping operators adjusted for a sample HDR image. The histogram in the background illustrates the luminance of the sample HDR image. Refer to "Comparison of Global Mapping Curves" for details

11.1.2 Spatially Variant Operators

A single pixel emitting radiance, which seems to be the smallest entity in computer graphics, is actually meaningless for the human eye. As obviously as the human eye cannot see a single pixel when all other pixels on the screen are of the same color, the HVS cannot retain much information from absolute luminance or radiance levels. Instead, we are perfectly equipped for the task of comparison of light patches – we see contrast rather than luminance. From this standpoint all spatially invariant tone mapping operators, which consider each pixel separately and transform its luminance using a monotonic tone reproduction curve, are poor in mimicking those important characteristics of the HVS. It turns out that global operators are also poor in reproducing local image details. To achieve better balance between local and global contrast reproduction, spatially variant tone mapping have been introduced.

Lightness is a perceptual quantity measured by the HVS which describes the amount of light reflected from the surface normalized for the illumination level. Contrary to brightness, which describes a visual sensation according to which an area exhibits more or less light, the lightness of a surface is judged relative to the brightness of a similarly illuminated area that appears to be white. The basic observation relevant for spatially variant operators is that the scene illumination is responsible for high dynamic range and must be compressed while lightness with surface details (texture) should be preserved. Since recovering lightness from images is an ill posed problem which can be solved precisely only for unrealistically simplified scenes (e.g., it is often assumed that illumination is a slowly changing, smooth function [13–15], which effectively means that shadows cannot be handled), approximated methods of lightness and illumination separation are used. For this purpose HDR images are decomposed into various layers in terms of spatial frequencies, which are then differently attenuated in terms of contrast. Low spatial frequency layers which supposedly carry more illumination information are attenuated stronger. Pyramids of band filters such as Laplacians or Gabor filters are used at the decomposition stage. Unfortunately, combining such differently attenuated layers into the output image leads to contrast reversal or halos [16]. This was a major problem with all early spatially variant operators.

In the following sections we briefly review modern local tone mapping operators which mostly avoid halo artifacts by considering sophisticated robust filters that penalize spatial neighbors (outliers) whose intensity is significantly different from the current pixel intensity ("Photographic Tone Reproduction", "Perceptually Uniform JND Space", and "Bilateral Filtering Method"). We present also a solution for handling difficult scenes featuring substantially different luminance levels and different chromatic adaptation conditions ("Lightness Perception"). In "Gradient methods" we discuss in more detail an operator featuring controllable enhancement of local and global image details (contrast), which belong to the class of gradient methods that rely on direct contrast processing (instead of luminance).

Photographic Tone Reproduction

Reinhard et al. presented a photographic tone reproduction inspired by photographic film development and the printing process [17]. The luminance of an image is initially compressed into the displayable range using a sigmoid function (refer to "Photoreceptor Model of Contrast Compression"). The photographic "dodging and burning" technique, which allows a different exposure for each part of the image, is used to enhance local contrast. To automate processes, low contrast regions are found by a center-surround filters at different scales. The biggest scale that does not contain outlier pixels is chosen. Then, a tone mapping function is locally applied. The automatic dodging and burning method enhances contrast and details in an image while preserving the overall luminance characteristics. This method can operate fully automatically based on information extracted individually from each HDR image, freeing the user from any parameter setting [18]. In Sect. 11.3 we provide more details on the photographic tone reproduction which is a good choice for real-time HDR video processing.

Perceptually Uniform JND Space

Ashikhmin presented a new, multipass approach to tone mapping [19]. The method takes into account two basic characteristics of the human visual systems (HVS): signaling absolute brightness and local contrast. This method first calculates local adaptation luminance by calculating an average luminance of neighboring pixels fitting in a bound-limited contrast range (similar to Reinhard et al. [17]). Then, it applies the capacity function which is based on the linearly approximated threshold vs. intensity function and calculates the final pixel values. The key feature of the capacity function is that it transforms input luminance values into uniformly scaled space in terms of just noticeable differences for various levels of adaptation luminance $\bar{Y}$. The final calculation restores the details which may be lost in the steps of compression. A tone mapped pixel value is obtained by multiplying a detail image given by the ratio of pixel luminance to the corresponding local world adaptation.

Bilateral Filtering Method

A fast bilateral filtering method was presented by Durand and Dorsey [20]. This method considers two different spatial frequency layers: a base layer and a detail layer. The base layer preserves high contrast edges and removes high-spatial frequency details of lower contrast. The detail layer is created as the difference of the original image and the base layer in logarithmic scale. After contrast compression in the base layer, both layers are summed up to create a tone mapped image.In Sect. 11.5 we provide more details on this operator, which is suitable for high-quality HDR video processing that is performed off-line.

Lightness Perception

Lightness constancy is an important characteristics of the HVS which leads to a similar appearance of the perceived objects independently of the lighting

images with areas illuminated by considerable different illuminants, as for example a shot in a room with tungsten illumination (orange color cast) with a window looking at the shadowed outdoor (blue color cast). The concept of frameworks enables also the application of the algorithm to video sequences where temporal coherence between frames can be achieved via the smoothing of the local anchors values.

Gradient Methods

Lightness determination algorithms [13–15] through recovering reflectance information naturally achieve dynamic range compression and are commonly used for tone mapping purposes [28]. In all those algorithms a logarithmically compressed intensity signal (an approximation of luminance processing by the retina) is differentiated and then thresholded to eliminate small derivative values attributed to illumination changes. The differentiation actually transforms images from the luminance to the contrast domain. The final integration of the thresholded derivatives leads to the reconstruction of image reflectance. In the Retinex algorithm [13] a huge number of global paths in the image plane is considered to perform such an integration. Horn [14] performs the differentiation and integration in 2-D, which can be reduced to the solution of Poisson's equation. Those algorithms rely on many limiting assumptions [15], such as slow changes in illumination, which are not very realistic for high contrast images. Fattal et al. [29] are the first who consider contrast processing explicitly for tone mapping. They show that visually compelling tone mapped images can be obtained through a smart nonlinear attenuation of luminance gradient magnitudes.

In this section we describe in more detail a more recent work of Mantiuk et al. [30] which can be seen as a generalization of [29]. Mantiuk et al. introduce a multi-resolution framework which extends contrast considerations from local pixel neighborhood to the whole image domain. In their approach the control over the global contrast (between distant image regions) and local contrast (between neighboring pixels) relies on the minimization problem rather than on user-tuned parameters as in [29]. This leads to images free from artifacts that are typical for local optimization procedures unable to grasp global contrast relations. Even without any picture-by-picture tuning, the framework can produce artifact-free and extremely sharp images.

An important role in the contrast domain framework of Mantiuk et al. play two distinct characteristics of the HVS: localized contrast perception and contrast discrimination threshold.

Localized contrast perception comes from the fact that the human eye can see high frequency patterns with foveal vision only within limited area, while the rest of the image is seen with parafoveal vision. For example, the eye can easily see the difference between fine details if they are close to each other, but have difficulties to distinguish which detail is brighter or darker if they are distant in the field of view. On the other hand, distant light patches can be easily compared if they are large enough. Based on those facts, Mantiuk et al. introduce a representation of contrast in complex images as a low-pass pyramid of contrast values. They use that pyramid to perform image processing on contrast rather than luminance.

The *contrast discrimination threshold* is the smallest difference in contrast the human eye can see. Such contrast discrimination data is well studied in psychophysics, but unfortunately mostly for low contrast patterns unsuitable for HDR images. The notable exception is the work of Whittle [31], in which the contrast discrimination was measured for the full range of contrast patterns. Whittle's data clearly show that the eye is the least sensitive (the thresholds are the highest) for low and high contrast ranges, in which either neural noise or optics of the eye limits our vision. Therefore, the contrast in those ranges can be modified the most before we notice any visible difference.

In the tone mapping algorithm [30], images are first converted to the multi-scale contrast pyramid. The contrast values are then modified accordingly to the contrast discrimination thresholds – the contrast is strongly compressed for very low and very high contrast and minimally compressed for the middle contrast range, to which the eye is the most sensitive. Another approach, which leads to even sharper but less natural looking images, takes advantage of the statistics of an image and performs histogram equalization on contrast values. The contrast values resulting from both approaches are then converted back to the luminance values by solving an optimization problem. Note that for the histogram equalization approach the basic difference with respect to [12] is that the latter one operates directly on luminance values while Mantiuk et al. consider contrast values instead.

Contrast domain offers several advantages over luminance domain. Firstly, contrast can be modified with regard to its magnitude, thus taking advantage of contrast perception characteristics such as the *contrast discrimination thresholds* becomes simple. Secondly, extremely sharp contrast can be achieved without introducing halo artifacts, since contrast values are compressed but not reversed. Finally, the pyramid of contrast values [30], which approximates the *localized contrast perception*, can take into account both local (high spatial frequency) and global (low frequency) contrast relations. Without taking into account global-contrast relations, the resulting images can show significantly disparate brightness levels in the areas where the input luminance is the same (refer to Fig. 11.3). Figure 11.4 shows the difference in the image sharpness achieved using the bilateral filtering approach [20], which is one of the best local tone mapping algorithms working in luminance domain, and the gradient domain technique as proposed by Mantiuk et al. [30].

11.2 HDR Video: Specific Conditions and Requirements

The tone mapping algorithms have been mostly developed specifically for the static images and they are not always applicable to the HDR video streams. The problems may among others arise from the temporal incoherence or too

Fig. 11.3. The algorithm by Fattal et al. (*left*) renders windows panes of different brightness due to the local nature of the optimization procedure. The contrast compression on the multiscale contrast pyramid proposed by Mantiuk et al. can maintain proper global contrast proportions (*right*). The approach based on the perceptual model of the contrast discrimination thresholds has been used to generate this image. The HDR source image courtesy of Greg Ward

Fig. 11.4. Extreme sharp contrast can be achieved without introducing halos. The result of bilateral filtering by Durand and Dorsey (*left*) compared with Mantiuk et al. method based on the histogram equalization (*right*). The HDR source image courtesy of Greg Ward

high computational complexity. In the following sections we focus on the tone mapping operators for video applications having particularly in mind the sensors described in the previous chapters of this book.

The capabilities of the HDR video cameras define specific characteristics of the video streams they generate. Currently the resolution is usually low (typically VGA), what implies that the tone mapping process may be done relatively fast or more complex methods can be used in the given time-slice between the frames. The dynamic range that the cameras are capable to capture is already very wide, therefore often an aggressive compression is necessary to fit the whole information into the displayable range. On the other hand the level of noise is still quite high especially during acquisition in poor illumination conditions. In this case, algorithms that extract details very well may cause the noise amplification and in fact lead to the loss of useful information.

Finally, the field of application for such HDR video cameras provides further requirements to the performance of the tone-mapping algorithms. Specific expectations to the obtained image-quality should be taken into account while choosing the algorithm.

11.3 Tone Mapping for HDR Video

In general, the global algorithms are usually the first choice for the tone mapping of HDR video. Mainly because of their simplicity and, in consequence, high efficiency, but also because it is straightforward to make them temporally coherent. What is more, often a type of dynamic range compression is already present in the camera response as is the case in for instance HDRC sensor which has a logarithmic response similar to the idea of logarithmic contrast compression ("Logarithmic Contrast Compression").

Sometimes however the dynamic range is so high that after the global tone mapping the colors in the image look desaturated and the fine details in the scene are lost. In such cases the local operators provide better results at the cost of the processing speed, however the real-time implementation of some of those algorithms is currently not achievable. As a compromise, various algorithms provide an enhancement to the global contrast compression which to some extent allows the preservation of details. In Sect. 11.3.1 we introduce the state-of-the-art solution in this category.

11.3.1 Response Curve Compression

The algorithm proposed by Reinhard et al. [17] operates solely on the luminance values which can be extracted from RGB intensities using the standard CIE XYZ transform (D65 white point). The method is composed of a global-scaling function similar to photoreceptor model ("Photoreceptor Model of Contrast Compression"), and a local dodging and burning technique which allows to preserve fine details. The results are driven by two parameters: the adapting luminance for the HDR scene and the key value. The adapting luminance ensures that the global scaling function provides the most efficient mapping of luminance to the display intensities for given illumination conditions in the HDR scene. The key value controls whether the tone mapped image appears relatively bright or relatively dark.

In this algorithm, the source luminance values Y are first mapped to the relative luminance Y_r:

$$Y_r = \frac{\alpha \cdot Y}{\bar{Y}}, \qquad (11.6)$$

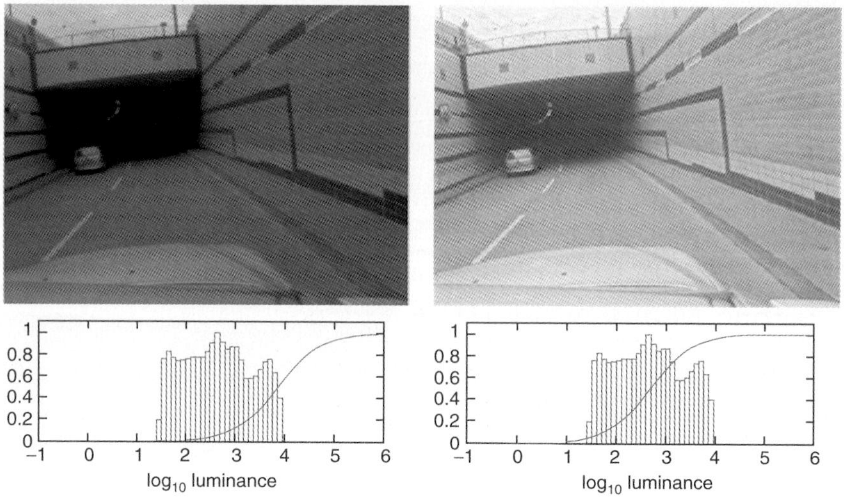

Fig. 11.5. Tone mapping of an HDR image with a low key (*left*) and a high key (*right*). The curve on the histograms illustrates how the luminance is mapped to normalized pixel intensities. Refer to Sect. 11.3 for details

where $\bar{Y}$ is the logarithmic average of the luminance in the scene, which is an approximation of the *adapting luminance*, and α is the key value. The relative luminance values are then mapped to the displayable pixel intensities L using the following function (compare to (11.2) in "Photoreceptor Model of Contrast Compression"):

$$L = \frac{Y_\mathrm{r}}{1 + Y_\mathrm{r}} \qquad (11.7)$$

The above formula maps all luminance values to the $[0:1]$ range in such way that the relative luminance $Y_\mathrm{r} = 1$ is mapped to the pixel intensity $L = 0.5$. This property is used to map a desired luminance level of the scene to the middle intensity on the display. Mapping higher luminance level to middle gray results in a subjectively dark image (low key) whereas mapping lower luminance to middle gray will give a bright result (high key) (see Fig. 11.5). Obviously, images which we perceive at night appear relatively dark compared to what we see during a day. We can simulate this impression by modulating the key value in (11.6) with respect to the adapting luminance in the scene. We explain our solution in Sect. 11.3.4.

11.3.2 Local Details Enhancement

Often, the tone mapping function in (11.7) may lead to the loss of fine details in the scene due to the extensive contrast compression. Reinhard et al. [17] propose a solution to preserve local details by employing a spatially variant local adaptation value V in (11.7):

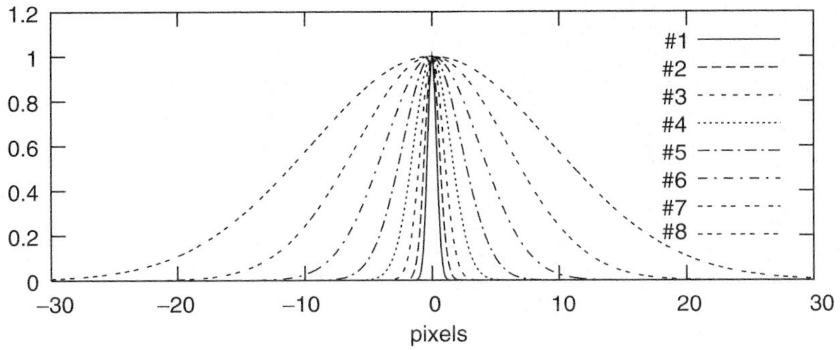

Fig. 11.6. Plot of the Gaussian profiles used to construct the scales of the pyramid used for local dodging and burning in the tone mapping algorithm. The smallest scale is #1 and the largest is #8. Plots are normalized by maximum value for illustration

$$L(x, y) = \frac{Y_{\mathrm{r}}(x, y)}{1 + V(x, y)}. \tag{11.8}$$

The local adaptation V equals to an average luminance in a surround of the pixel. The problem lies however in the estimation of how large the surround of the pixel should be. The goal is to have as wide surround as possible, however too large area may lead to well known inverse gradient artifacts, *halos*. The solution is to successively increase the size of a surround, checking each time if no artifacts are introduced. For this purpose a Gaussian pyramid is constructed with successively increasing kernel

$$g(x, y, s) = \frac{1}{\pi s^2} \cdot \mathrm{e}^{-\frac{x^2+y^2}{s^2}}. \tag{11.9}$$

The Gaussian for the first scale is one pixel wide, setting kernel size to $s = (2\sqrt{2})^{-1}$, on each following scale s is 1.6 times larger. The Gaussian functions used to construct seven scales of the pyramid are plotted in Fig. 11.6. As we later show, such a pyramid is also very useful in introducing the perceptual effects to tone mapping.

11.3.3 Temporal Luminance Adaptation

While tone mapping the sequence of HDR frames, it is important to note that the luminance conditions can significantly change from frame to frame. The human vision reacts to such changes through the temporal adaptation processes. The time course of adaptation differs depending on whether we adapt to light or to darkness, and whether we perceive mainly using rods (during night) or cones (during a day). Many intricate models have been introduced to computer graphics, however it is not as important to faithfully model the process as to account for it at all [32].

In the tone mapping algorithm chosen by us, the luminance adaptation can be modeled using the adapting luminance term in (11.6). Instead of using the actual adapting luminance $\bar{Y}$, a filtered value $\bar{Y}_{\mathrm{a}}$ can be used whose value changes according to the adaptation processes in human vision, eventually reaching the actual value if the adapting luminance is stable for some time. The process of adaptation can be modeled using an exponential decay function [33]

$$\bar{Y}_{\mathrm{a}}^{\mathrm{new}} = \bar{Y}_{\mathrm{a}} + (\bar{Y} - \bar{Y}_{\mathrm{a}}) \cdot (1 - \mathrm{e}^{-\frac{T}{\tau}}), \tag{11.10}$$

where T is the discrete time step between the display of two frames, and τ is the time constant describing the speed of the adaptation process. The time constant is different for rods and for cones

$$\tau_{\mathrm{rods}} = 0.4\mathrm{s} \quad \tau_{\mathrm{cones}} = 0.1\mathrm{s}. \tag{11.11}$$

Therefore, the speed of the adaptation depends on the level of the illumination in the scene. The time required to reach the fully adapted state depends also whether the observer is adapting to light or dark conditions. The values in (11.11) describe the adaptation to light. For practical reasons the adaptation to dark is not simulated because the full process takes up to tens of minutes. Therefore, it is acceptable to perform the adaptation symmetrically, neglecting the case of a longer adaptation to dark conditions.

We model the temporal luminance adaptation based on (11.10). However, in our algorithm we do not perform separate computations for rods and cones, what makes it hard to properly estimate the adaptation speed having two time constants τ_{rod} and τ_{cone} instead of one. To account for this, and still be able to correctly reproduce the speed of the adaptation, we interpolate the actual value of the time constant based on the sensitivity of rods (11.15):

$$\tau(\bar{Y}) = \sigma(\bar{Y}) \cdot \tau_{\mathrm{rod}} + (1 - \sigma(\bar{Y})) \cdot \tau_{\mathrm{cone}} \tag{11.12}$$

which we then use to process the adaptation value using (11.10).

11.3.4 Key Value

The key value, explained in Sect. 11.3.1, determines whether the tone-mapped image appears relatively bright or dark, and in the original paper [17] is left as a user choice. In his follow-up paper, Reinhard [18] proposes a method of automatic estimation of the key value that is based on the relations between minimum, maximum, and average luminance in the scene. Although the results are appealing, we feel this solution does not necessary correspond to the impressions of everyday perception. The critical changes in the absolute luminance values may not always affect the relation between the three values. This may lead to dark night scenes appearing too bright and very light too dark.

The key value, α in (11.6), takes values from $[0:1]$ range where 0.05 is the low key, 0.18 is a typical choice for moderate illumination, and 0.8 is

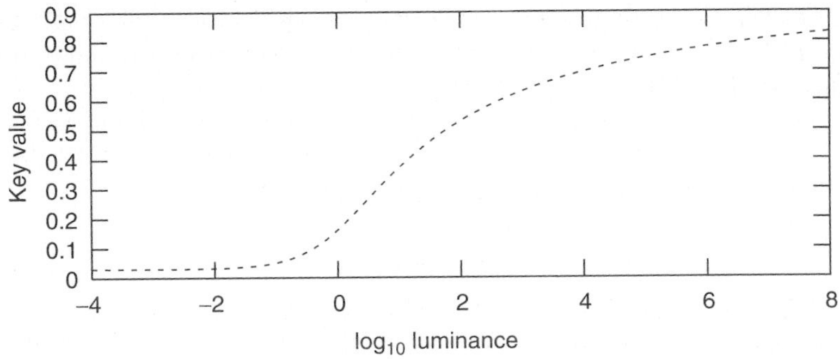

Fig. 11.7. Key value related to adapting luminance in the scene

the high key. We propose to calculate the key value based on the absolute luminance. Since the key value has been introduced in photography, there is no scientifically-based experimental data which would provide an appropriate relation between the key value and the luminance, so the proper choice is a matter of experience. We therefore empirically specify key values for several illumination conditions and interpolate the rest using the following formula:

$$\alpha(\bar{Y}) = 1.03 - \frac{2}{2 + \log_{10}(\bar{Y} + 1)}, \tag{11.13}$$

where α is the key value and $\bar{Y}$ is an approximation of the adapting luminance. The plot of this estimation is shown in Fig. 11.7.

11.3.5 Tone Mapping

We start the tone mapping process with calculating the luminance from the HDR frame and mapping it to the relative luminance according to (11.6). We calculate the logarithmic average of the luminance $\bar{Y}$ in the frame and apply the temporal adaptation process (11.10). The map of relative luminance values constitutes the first scale of the Gaussian pyramid. At each scale of the Gaussian pyramid, we render the successive scale by convolving the previous scale with the appropriate Gaussian (11.9). Having the current and the previous scales, the local adaptation is computed using the measure of the difference between the previous and the current scale as described in [17]. We iterate through all of the scales of the Gaussian pyramid and in the final iteration we fill the missing parts of the adaptation map with data from the lowest scale.

We then use the following formula to calculate the tone-mapped RGB values of the original HDR input

$$\begin{bmatrix} R_L \\ G_L \\ B_L \end{bmatrix} = \begin{bmatrix} R \\ G \\ B \end{bmatrix} \cdot \frac{L}{Y}, \tag{11.14}$$

where $\{R_L, G_L, B_L\}$ denotes the tone mapped intensities, $\{R, G, B\}$ are the original HDR values, Y is the luminance, and L is the tone mapped luminance calculated using (11.8) from Sect. 11.3.1.

11.4 Simulating Perceptual Effects

The fact that a daylight scene appears very bright and colorful, but during the night everything looks dark and grayish, is obvious for an average observer. In dimly illuminated scenes the fine details are not perceivable, because the acuity of human vision is degraded, therefore when using local tone mapping operators, well preserved details will give an unrealistic impression in such cases. On the other hand, certain perceptual effects, like glare, cannot be evoked because the maximum luminance of typical displays is not high enough. However, we are so used to the presence of such a phenomena, that adding glare to an image can increase subjective brightness of the tone mapped image [34, 35]. Apparently, it appears important to properly predict and simulate such perceptual effects during the tone mapping process in order to convey a realistic impression of HDR data over a wide range of luminance, when such data are displayed on typical display devices.

The simulation of the perceptual effects is in principle possible, because the camera response can be calibrated to approximate the physically correct values $\mathrm{cd{\cdot}m^{-2}}$ at given aperture [26]. Knowing exactly the illumination level in the scene, it is possible to estimate the performance of HVS in this conditions. In the following sections, we present the models which describe the most significant perceptual effects and explain how to combine them with the tone-mapping algorithm given in Sect. 11.3.1.

11.4.1 Scotopic Vision

The human vision operates in three distinct adaptation conditions: scotopic, mesopic, and photopic. The photopic and mesopic vision provide color vision, however in scotopic range, where only rods are active, color discrimination is not possible. The cones start to loose their sensitivity at $3.4\,\mathrm{cd{\cdot}m^{-2}}$ and become completely insensitive at $0.03\,\mathrm{cd{\cdot}m^{-2}}$ where the rods are dominant. We model the sensitivity of rods σ after [36] with the following function:

$$\sigma(Y) = \frac{0.04}{0.04 + Y}, \tag{11.15}$$

where Y denotes the luminance. The sensitivity value $\sigma = 1$ describes the perception using rods only (monochromatic vision) and $\sigma = 0$ perception using cones only (full color discrimination). The plot of (11.15) is shown in Fig. 11.8.

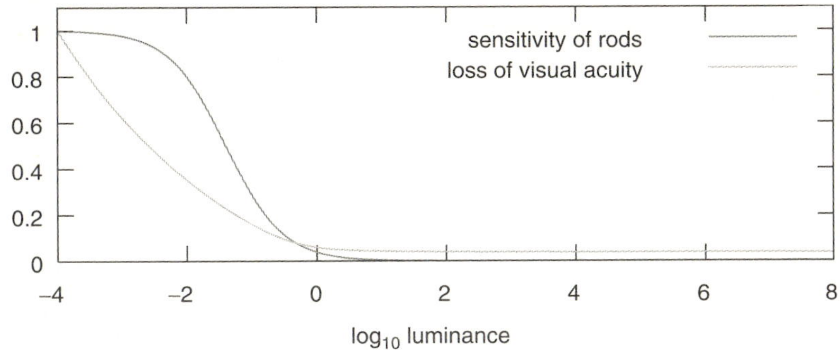

Fig. 11.8. The influence of perceptual effects on vision depending on the luminance level. For details on rods sensitivity and visual acuity refer to Sects. 11.4.1 and 11.4.2, respectively

11.4.2 Visual Acuity

Perception of spatial details in the human vision is not perfect and becomes limited with a decreasing illumination level. The performance of visual acuity is defined by the highest resolvable spatial frequency and has been investigated by Shaler in [37]. Ward et al. [12] offer the following function fit to the data provided by Shaler

$$\mathrm{RF}(Y) = 17.25 \cdot \arctan(1.4 \log_{10} Y + 0.35) + 25.72, \qquad (11.16)$$

where Y denotes the luminance and RF is the highest resolvable spatial frequency in cycles per degree of the visual angle. The plot of this function is shown in Fig. 11.9.

To simulate the loss of visual acuity on a display device we need to map the visual degrees to pixels. Such a mapping depends on the size of the display, the resolution, and the viewing distance. For a typical observation of a 15-inch screen from half a meter at 1024×768 resolution we assume 45 pixels per 1 degree of the visual angle. It is important to note that the highest frequency possible to visualize in such conditions is 22 cycles per visual degree. Therefore, technically we can simulate the loss of visual acuity only for luminance below $0.5\,\mathrm{cd}\cdot\mathrm{m}^{-2}$. The irresolvable details can be removed from an image by the convolution with the Gaussian kernel from (11.9) where s is calculated as follows [12]:

$$s_{\mathrm{acuity}}(Y) = \frac{\mathrm{width}}{\mathrm{fov}} \cdot \frac{1}{1.86 \cdot \mathrm{RF}(Y)}, \qquad (11.17)$$

where "width" denotes the width in pixels and "fov" is the horizontal field of view in visual degrees. For typical observation the "width" to "fov" relation equals 45 pixels.

The resulting tone mapped images feature good balance between global contrast and detail visibility, which can be easily adjusted accordingly to the user preferences. The method does not lead to halo artifacts and reversing contrast, which are common drawbacks for many local tone mapping operators.

The bilateral tone mapping method is not intended to model the HVS and can be considered as a purely image processing method. In this sense this tone mapping method does not comply with the definition formulated by Tumblin and Rushmeier [3] which states that the basic goal of a tone reproduction operator is to achieve a good match between the appearance of real world scenes and corresponding tone mapped images.

Figure 11.12 shows a comparison of an HDR image recorded with an HDRC video camera. The colors of the stretched logarithmic data on the left are washed out, details are moderately preserved. On the right side, the image of the eye using bilateral filter technique is shown. Note that the fine details of the iris, skin, or the blood vessels are well preserved.

Choudhury and Tumblin [41] introduced trilateral filtering, which removes some of bilateral filtering drawbacks such as blending between disjoint regions that have similar intensity characteristics and smoothing high frequency details in high gradient regions. However, those improvements come at the expense of significant computational costs.

The HDRC sensor technology (refer to Chap. 2) with its logarithmic response curve makes the use of the bilateral filter technique interesting and easy because all of the image calculations are done in the logarithmic space.

Durand and Dorsey tested their bilateral filter tone mapping mostly for still images of very high quality. The operator has not been applied to dynamic sequences, in particular, those featuring significant temporal noise as this is the case for currently available HDR video cameras.

Typically, HDR images are created by fusing a set of photographs taken with different exposures using a standard LDR camera featuring linear sensor (for more details refer to the description of multiple-exposure techniques in Chap. 13). Image artifacts like ghosting due to object movements, lens flare can be removed or limited with available postprocessing algorithms. Such computed HDR images feature low dark noise as well as good contrast and color saturation, which are perfect initial conditions for all tone mapping algorithms to achieve good visual results.

On the other side, HDR video cameras like the HDRC camera enable direct capturing of live HDR video streams (see Fig. 11.13). A single frame of the video stream is usually not so "ideal" as an image recorded with a multiple-exposure method, which means that they include different kinds of noise like fixed-pattern or readout noise. Also artifacts from camera components appear like ghosts created by reflections of the optical system or additional filters during fluctuating recording environments (e.g., car driving scenes), blurred

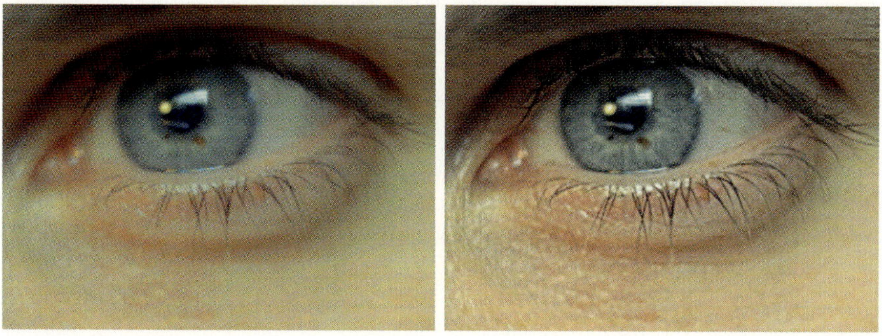

Fig. 11.12. HDRC image with stretched log mapping (*left*) compared with bilateral filtering mapping (*right*) (Full dynamic range of the images is not visualized due to the print limitations)

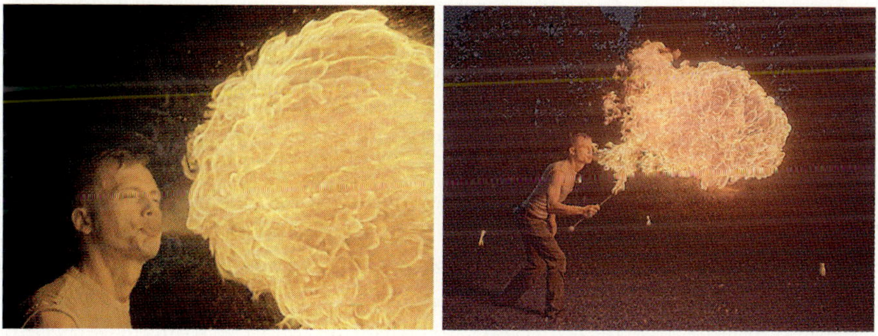

Fig. 11.13. Images from HDRC sensor of a fire breather movie with tone mapping using bilateral filtering. This type of sensor becomes more and more interesting for automotive, machine vision, surveillance systems, and television applications

Fig. 11.14. HDRC image with stretched log mapping (*left*) compared with Bilateral Filtering mapping (*right*)

image regions caused by not completely IR-filtered images (e.g., high beam car headlights that produce their lighting power peak in the NIR range, see Fig. 11.14) conditions that cannot be changed during recording. All these artifacts cause the loss of detailed image information and will also complicate the use of a tone mapping algorithm.

So it is necessary to preprocess the image data to get a better image quality for tone mapping functions. In the case of detail preserving operators, any image interferences like noise, reflections from the optics will be preserved as an image detail and amplified.

In practice, the following settings for the bilateral tone mapping operator worked well for many HDRC applications: $\sigma_S = 4$, $\sigma_I = 0.4$, and the compression ratio c applied to the base layer ranging from 0.3 to 0.5.

It is also recommended to reduce temporal noise first (refer to Chap. 3), e.g., using a simple 1D temporal filtering with motion compensation as proposed by Bennett and McMillan [42].

A relevant issue is the correction of the sensor's fixed pattern noise (discussed in Sect. 2.3). The colors look unsaturated after color demosaicking due to the logarithmic principle of lightness recording, so it is important to enhance additionally the color saturation (discussed in Chap. 4). Enhancement factors for $\alpha_R = \alpha_G = \alpha_B = 1.3$–$1.5$ worked well in conjunction with bilateral tone mapping.

In summary, the local tone mapping operator provides good overall brightness and contrast especially for images with very HDR in comparison to other global operators. This means that it is possible to display nearly the whole dynamic range of an HDRC camera image without loss of image details on an LDR display and fully demonstrate the power of the logarithmic compression concept of HDRC technology.

Fig. 11.14 shows an HDR scene recorded with an HDRC camera of high-beam car headlights against dark backlight illumination with a dynamic range of more than six orders of magnitude (illumination range varied from 0.3 lx to over 300,000 lx). The colors of the stretched logarithmic compressed image (Fig. 11.14, left) look unsaturated and many image details of the scene e.g., the Macbeth Color Chart or the LED lamp on the bottom left of the image are lost. In comparison to the bilateral filter mapped image (Fig. 11.14, right) all colors of the color checker are pleasantly saturated and fine details are visible.

Spatially varying filtering including bilateral filtering is a very time consuming operation. Durand and Dorsey [20] proposed an approximated bilateral filtering solution, which reduces the computation time significantly, but still real-time performance as required for on-line HDR video playback cannot be at present achieved.

For applications involving real-time HDR video playback a very good cost-performance can be achieved using commodity graphics cards. It can be expected that with the increasing power and programmability of graphics processing units (GPUs), more and more advanced image processing and filtering will be feasible in the near future.

11.6 Summary

In view of the increasing availability of the HDR contents the problem of their presentation on conventional display devices is highly recognized. Different goals and approaches led to the development of versatile algorithms. These algorithms have different properties which correspond to the specific requirements and applications. Furthermore, due to the temporal incoherence certain methods cannot be used for the tone mapping of video streams. A universal method has not been found so far, therefore the choice of the tone-mapping method should be based on the application requirements.

With respect to the HDR video streams the choice of an appropriate tone mapping method is usually a trade-off between the computational intensity and the quality of dynamic range compression. The quality here is mainly assessed by a good local details visibility. The global tone mapping methods are very fast, but often lead to the loss of local details due to an intensive dynamic range compression. Such methods should be used whenever high efficiency is the main requirement of the target application. The adaptation mechanisms can be used to select the range of luminance values which should obtain the best mapping. However when the quality is insufficient, local tone mapping methods are necessary. The local details enhancement methods provide a good improvement to the global tone mapping methods still achieving good computational performance. The best results are achieved using the bilateral filtering, lightness perception or gradient methods, however these techniques do not currently have real-time implementations. They can be therefore used for an off-line rendering of high quality still shots.

The photometrically calibrated HDR video streams allow for the prediction of the perceptual effects. Such effects are typical to everyday perception of real-world scene, but do not appear when observing a display showing a tone mapped HDR video. Prediction of such effects and their simulation can increase the realism of the presentation of HDR contents. On the other hand, such a prediction may also be used to identify situations when a real-world observation of scene would be impaired and to hint the tone mapping algorithm to focus on the good detail reproduction there.

References

1. H. Seetzen, W. Heidrich, W. Stuerzlinger, G. Ward, L. Whitehead, M. Trentacoste, A. Ghosh, and A. Vorozcovs. High dynamic range display systems. *ACM Transactions on Graphics*, 23(3), 2004
2. B.A. Wandell. *Foundations of Vision.* Sinauer Associates, Sunderland, MA, 1995

3. J. Tumblin and H.E. Rushmeier. Tone reproduction for realistic images. *IEEE Computer Graphics and Applications*, 13(6):42–48, November 1993
4. G. Ward. A contrast-based scalefactor for luminance display. *Graphics Gems IV*, pages 415–421, 1994
5. J.A. Ferwerda, S. Pattanaik, P.S. Shirley, and D.P. Greenberg. A model of visual adaptation for realistic image synthesis. In *Proceedings of SIGGRAPH 96*, Computer Graphics Proceedings, Annual Conference Series, pages 249–258, August 1996
6. S.N. Pattanaik, J.E. Tumblin, H.Yee, and D.P. Greenberg. Time-dependent visual adaptation for realistic image display. In *Proceedings of ACM SIGGRAPH 2000*, Computer Graphics Proceedings, Annual Conference Series, pages 47–54, July 2000
7. E. Reinhard and K. Devlin. Dynamic range reduction inspired by photoreceptor physiology. *IEEE Transactions on Visualization and Computer Graphics*, 11(1):13–24, 2005
8. T.G. Stockham. Image processing in the context of a visual model. *Proceedings of the IEEE*, 13(6):828–842, 1960
9. F. Drago, K. Myszkowski, T. Annen, and N. Chiba. Adaptive logarithmic mapping for displaying high contrast scenes. *Computer Graphics Forum, Proceedings of Eurographics 2003*, 22(3):419–426, 2003
10. K. Perlin and E.M. Hoffert. Hypertexture. In *Computer Graphics (Proceedings of SIGGRAPH 89)*, Vol. 23, pages 253–262, July 1989.
11. A. Yoshida, V. Blanz, K. Myszkowski, and H.P. Seidel. Perceptual evaluation of tone mapping operators with real-world scenes. In B.E. Rogowitz, T.N. Pappas, and S.J. Daly, editors, *Human Vision and Electronic Imaging X, IS&T/SPIE's 17th Annual Symposium on Electronic Imaging (2005)*, volume 5666 of *SPIE Proceedings Series*, pages 192–203, San Jose, USA, 2005. SPIE
12. G. Ward Larson, H. Rushmeier, and C. Piatko. A visibility matching tone reproduction operator for high dynamic range scenes. *IEEE Transactions on Visualization and Computer Graphics*, 3(4):291–306, 1997
13. E.H. Land and J.J. McCann. Lightness and the retinex theory. *Journal of the Optical Society of America*, 61(1):1–11, 1971
14. B.K.P. Horn. Determining lightness from an image. *Computer Graphics and Image Processing*, 3(1):277–299, 1974
15. A. Hurlbert. Formal connections between lightness algorithms. *Journal of the Optical Society of America A*, 3(10):1684–1693, 1986
16. J. DiCarlo and B. Wandell. Rendering high dynamic range images. In *Proceedings of SPIE*, Vol. 3965, pages 392–401, 2000
17. E. Reinhard, M. Stark, P. Shirley, and J. Ferwerda. Photographic tone reproduction for digital images. *ACM Transactions on Graphics*, 21(3):267–276, 2002
18. E. Reinhard. Parameter estimation for photographic tone reproduction. *Journal of Graphics Tools*, 7(1):45–52, 2002
19. M. Ashikhmin. A tone mapping algorithm for high contrast images. In *Proceedings of the 13th Eurographics workshop on Rendering*, pages 145–156, 2002
20. F. Durand and J. Dorsey. Fast bilateral filtering for the display of high-dynamic-range images. *ACM Transactions on Graphics*, 21(3):257–266, 2002
21. S.E. Palmer. *Vision Science: Photons to Phenomenology*, chapter 3.3 Surface-Based Color Processing. MIT, Cambridge, MA, 1999
22. A. Gilchrist. Lightness contrast and failures of constancy: A common explanation. *Perception & Psychophysics*, 43:415–424, 1988

23. I. Rock. *The Logic of Perception.* MIT, Cambridge, MA, 1983
24. A. Gilchrist and J. Cataliotti. Anchoring of surface lightness with multpile illumination levels. *Investigative Ophthamalmology and Visual Science*, 35, 1994
25. A. Gilchrist, C. Kossyfidis, F. Bonato, T. Agostini, J. Cataliotti, X. Li, B. Spehar, V. Annan, and E. Economou. An anchoring theory of lightness perception. *Psychological Review*, 106(4):795–834, 1999
26. G. Krawczyk, M. Goesele, and H.P. Seidel. Photometric calibration of high dynamic range cameras. Research Report MPI-I-2005-4-005, Max-Planck-Institut für Informatik, Stuhlsatzenhausweg 85, 66123 Saarbrücken, Germany, April 2005
27. G. Krawczyk, K. Myszkowski, and H.P. Seidel. Lightness perception in tone reproduction for high dynamic range images. In *The European Association for Computer Graphics 26th Annual Conference EUROGRAPHICS 2005*, volume 24 of *Computer Graphics Forum*, pages 635–646, Dublin, Ireland, 2005. Blackwell
28. D.J. Jobson, Z. Rahman, and G.A. Woodell. Properties and performance of a center/surround retinex. *IEEE Transactions on Image Processing*, 6(3):451–462, 1997
29. R. Fattal, D. Lischinski, and M. Werman. Gradient domain high dynamic range compression. *ACM Transactions on Graphics*, 21(3):249–256, 2002
30. R. Mantiuk, K. Myszkowski, and H.P. Seidel. A Perceptual framework for contrast processing of high dynamic range images. In J.Koenderink and J. Malik, editors, *Proceedings of the 2nd Symposium on Applied Perception in Graphics and Visualization (APGV 2005)*. ACM, New York, 2005
31. P. Whittle. Increments and decrements: luminance discrimination. *Vision Research*, 26(10):1677–1691, 1986
32. N. Goodnight, R. Wang, C. Woolley, and G. Humphreys. Interactive time-dependent tone mapping using programmable graphics hardware. In *Rendering Techniques 2003: 14th Eurographics Symposium on Rendering*, pages 26–37, 2003
33. F. Durand and J. Dorsey. Interactive tone mapping. In *Rendering Techniques 2000: 11th Eurographics Workshop on Rendering*, pages 219–230, 2000
34. G. Spencer, P. Shirley, K. Zimmerman, and D.P. Greenberg. Physically-based glare effects for digital images. In *Proceedings of ACM SIGGRAPH 95*, pages 325–334, 1995
35. G. Spencer, P.S. Shirley, K. Zimmerman, and D.P. Greenberg. Physically-based glare effects for digital images. In *Proceedings of ACM SIGGRAPH 95*, Computer Graphics Proceedings, Annual Conference Series, pages 325–334, August 1995
36. R.W.G. Hunt. *The Reproduction of Colour in Photography, Printing and Television: 5th Edition.* Fountain, Tolworth 1995
37. S. Shaler. The relation between visual acuity and illumination. *Journal of General Psychology*, 21:165–188, 1937
38. R.J. Deeley, N. Drasdo, and W. N. Charman. A simple parametric model of the human ocular modulation transfer function. *Ophthalmology and Physiological Optics*, 11:91–93, 1991
39. C. Tomasi and R. Manduchi. Bilateral filtering for gray and color images. In *Proceedings of the 1998 IEEE International Conference on Computer Vision*, Bombay, India, pages 839–846, 1998

40. J. Tumblin and G. Turk. LCIS: A boundary hierarchy for detail-preserving contrast reduction. In *Proceedings of ACM SIGGRAPH 99*, Computer Graphics Proceedings, Annual Conference Series, pages 83–90, 1999
41. P. Choudhury and J. Tumblin. The trilateral filter for high contrast images and meshes. In *Rendering Techniques 2003: 14th Eurographics Symposium on Rendering*, pages 186–196, 2003
42. E.P. Bennett and L. McMillan. Video enhancement using per-pixel virtual exposures. *ACM Transactions on Graphics*, 24(3): 765–776, 2005

12

HDR Image and Video Compression

Rafał Mantiuk

12.1 Introduction

HDR data can consume enormous amounts of storage space or transmission bandwidth if it is represented in a naive way. HDR compression can reduce the size of visual data by more than several hundreds of times without introducing objectionable distortions. Therefore, compression has become an important part of many HDR image and video systems.

Capturing of HDR video and images becomes easier with the development of HDR cameras. On the other end of the pipeline, HDR data can be displayed on new generation of HDR displays. However, in order to make those two ends of the pipeline work together, there is a need for a common format of data that both would understand. This can be achieved with *scene-referred* approach, described in Sect. 12.2.

The recent advances in digital camera and display technologies make a traditional 8-bit per color channel representation of visual data insufficient. The usual pixel formats as well as file formats for images and video (JPEG, PNG, MPEG) fail to represent scenes of dynamic range over 2 or 3 orders of magnitude and extended color gamut.

To address this problem, several HDR image formats have been introduced. Section 12.4 describes several such image formats, which are the most commonly used. Equally important to image compression is the compression of HDR video. Therefore, Sect. 12.5 describes a HDR extension to MPEG encoding. The major distinction between low dynamic range and HDR compression is the format in which pixels are stored. A pixel format, which can encode the full range of color the eye can see and that is adjusted to the limitations of our perception, is described in Sect. 12.6. Finally, Sect. 12.7 gives a brief review of the existing software for HDR image and video processing.

12.2 Device-Referred and Scene-Referred Representation of Images

Commonly used image formats (JPEG, PNG, TIFF, etc.) contain data that is tailored to particular display devices: cameras, CRTs or LCD monitors. The sRGB color space, a de facto standard for many low dynamic images and devices, defines the maximum luminance of 80 $\mathrm{cd\,m^{-2}}$ and a restricted color gamut, which does not even cover the gamut of some color printers. Obviously, such limited representation of color cannot faithfully represent real world scenes, which can contain both highly saturated colors and luminance range that cannot be captured by any low-dynamic range color space. The reason why sRGB is so restrictive is that it was designed to match the capabilities of most computer monitors. Obviously, such representation of images vaguely relates to the actual photometric properties of the scene it depicts, but it is rather dependent on a display device. Therefore, sRGB and similar color formats can be considered as device-referred (also known as output-referred) since they are tightly coupled with the capabilities and characteristics of a particular imaging device.

Because both color gamut and dynamic range may differ significantly between devices, there is often a need to convert image colors between different device-referred formats. ICC color profiles [1] can be used to convert visual data from one device-referred format to another. Such profiles define colorimetric properties of a device the image is intended for. Problems arise if the two devices have a different color gamut or dynamic range, in which case a conversion from one format to another usually requires loss of some visual information. The problem is even more difficult when an image captured with a HDR camera is to be converted to the color space of a low-dynamic range monitor (this is known as a tone mapping problem, discussed in Chapter 11 of this book). Obviously, the ICC profiles cannot be easily used to facilitate interchange of data between LDR and HDR devices.

A scene-referred representation of images offers a much simpler solution to this problem. It implies that an image should encode the actual photometric characteristic of the scene it depicts. Conversion from such a common representation (for example CIE XYZ tri-stimulus values), which directly corresponds to the physical luminance or radiance values, to a format suitable for a particular device is a responsibility of that device. HDR file formats are an example of scene-referred encoding, as they usually represent either luminance or radiance, rather than gamma corrected intensity.

12.3 HDR Image and Video Compression Pipeline

Although image and video compression of low-dynamic range sceneries has been studied thoroughly for many years, the problem of HDR compression has been identified and addressed only recently. Fortunately, most of low-dynamic

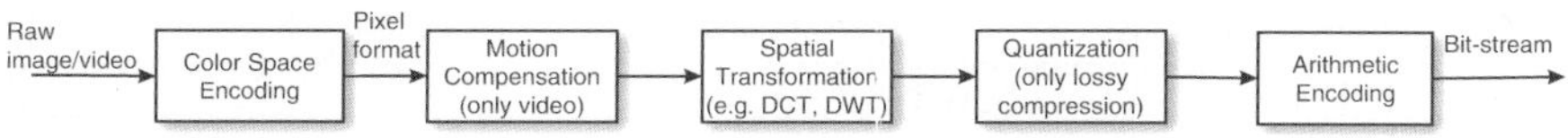

Fig. 12.1. A general pipeline of video and image compression. See the text for details

range compression techniques can also be applied to HDR with some minor modifications. In this chapter we focus on the differences between high- and low-dynamic range image and video compression without going too much into details of general compression techniques.

Figure 12.1 illustrates the basic pipeline of almost any video and image compression method. The first step is *color space encoding*, which aims at converting pixel values to the representation, which is more suitable for compression. This is usually the step that differs significantly between low-dynamic range and HDR compression. The next step, *motion compensation*, is present only in the case of video compression and is responsible for minimizing the differences between two consecutive frames due to motion, so that the difference between these frames can be encoded efficiently. Advanced compression methods do not store pixels, but rather transform images or video to the frequency or the wavelet domain, which gives much more compact representation (step *Spatial Transformation*). To achieve high compression ratios, the data is then quantized (step *Quantization*), which means that the precision and, thus, the number of bits needed to encode values is reduced. The final step, *Arithmetic encoding*, applies lossless compression to the data in order to further reduce its size.

Since the major difference between low-dynamic range and HDR compression is the way pixels are encoded, the focus of Sect. 12.4 is on pixel formats (or color spaces) used for HDR image compression. A detailed discussion of a particular HDR pixel format, which takes into account the limitations of the human eye, is described in Sect. 12.6.

12.4 HDR Image Formats

This section reviews the most popular scene-referred HDR image formats and gives some insight how HDR data can be efficiently represented and stored.

12.4.1 Radiance's HDR Format

One of the first HDR image formats, which gained much popularity, was introduced with the Radiance rendering package.[1] Therefore, it is known as the *Radiance* picture format [2] and can be recognized by the file extensions

[1] Radiance is an open source light simulation and realistic rendering package. Home page: `http://radsite.lbl.gov/radiance/`

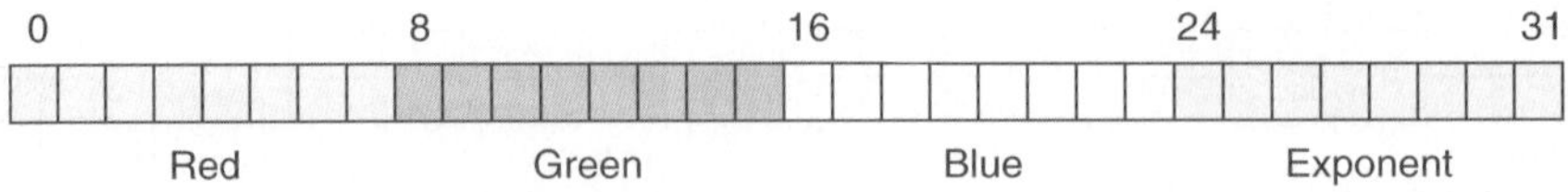

Fig. 12.2. 32-bit per pixel RGBE encoding

.hdr or .pic. The file consist of a short text header, followed by run-length encoded pixels. The pixels are encoded using so-called *RGBE* or *XYZE* representations, which differ only by a color space that is used. Since both representations are very similar, we only describe the *RGBE* encoding.

RGBE pixel encoding represents colors using four bytes: the first three bytes encode red, green and blue color channels, and the last byte is a common exponent for all channels (see Fig. 12.2). RGBE is essentially a custom floating point representation of pixel values. RGBE encoding takes advantage of the fact that all color channels are strongly correlated in RGB color space and their values are at least of the same order of magnitude. Therefore, there is no need to store a separate exponent for each color channel.

12.4.2 LogLuv TIFF

The major drawback of floating point representation of pixel values is that floating point numbers do not compress well. This is mainly because they need additional bits to encode both mantissa and exponent values. Such representation, although flexible, is not really required for visual data. Furthermore, precision error of floating point numbers varies significantly across the full range of possible values and is different than the "precision" of our visual system. Therefore, better compression can be achieved when integer numbers are used to encode HDR pixels.

The *LogLuv* encoding [3] requires only integer numbers to encode a full range of luminance and color gamut that is visible to the human eye. It is an optional encoding in the TIFF library. This encoding benefits from the fact that the human eye is not equally sensitive to all luminance ranges. In the dark we can see a luminance difference of a few hundredths of $\mathrm{cd\,m^{-2}}$, while in the sunlight we need a difference of tens of $\mathrm{cd\,m^{-2}}$ to see a difference. But if, instead of luminance, a logarithm of luminance is considered, the detectable threshold values do not vary so much and a constant value can be a conservative approximation of the visible threshold. Therefore, if a logarithm of luminance is encoded using integer numbers, quantization errors roughly correspond to the visibility thresholds of the human visual system, which is a desirable property for pixel encoding.

32-bit LogLuv encoding uses two bytes to encode luminance and another two bytes to represent chrominance (see Fig. 12.3). Chrominance is encoded using a perceptually uniform chromacity scale u' v' (see Sect. 12.6 for details). There is also 24-bit LogLuv encoding, which needs fewer bits to encode pixels

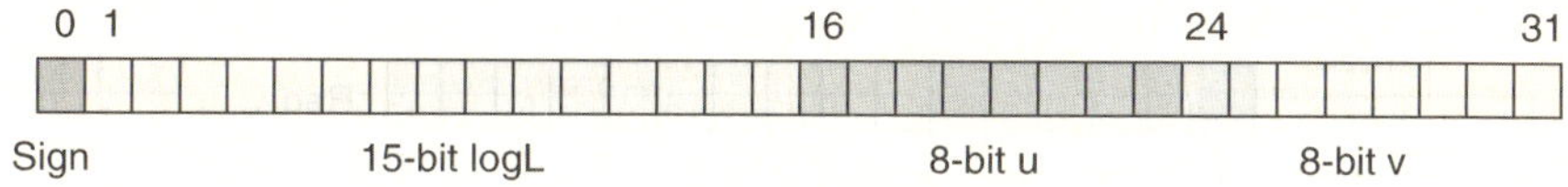

Fig. 12.3. 32-bit per pixel LogLuv encoding

with the precision that is below the visibility thresholds. However, this format is rather ineffective to encode, due to discontinuities resulting from encoding two chrominance channels with a single lookup value.

12.4.3 OpenEXR

An OpenEXR format or the EXtended Range format recognized by the file name extension .exr was made available with an open source C++ library in 2002 by Industrial Light and Magic (see `http://www.openexr.org/` and [4]). Before that date the format was used internally by Industrial Light and Magic for the purpose of a special effect production. The format is currently promoted as a special-effect industry standard and many software packages already support it. Some features of this format include:

- Support for 16-bit floating point, 32-bit floating point, and 32-bit integer pixels. The 16-bit floating point format, called "half", is compatible to the *HALF* data type in NVIDIA's Cg graphics language and supported natively on their new GeForce FX and Quadro FX 3D graphics solutions.
- Multiple lossless image compression algorithms. Some of the included codecs can achieve 2:1 lossless compression ratios on images with film grain.
- Extensibility. New compression codecs and image types can easily be added by extending the C++ classes included in the OpenEXR software distribution. New image attributes (strings, vectors, integers, etc.) can be added to OpenEXR image headers without affecting backward compatibility with existing OpenEXR applications.

Although OpenEXR file format offers several data types to encode channels, color data is usually encoded with 16-bit floating point numbers, known as half-precision floating points. Such two byte floating point number consists of one bit of sign, 5-bit exponent, and 10-bit mantissa, as shown in Fig. 12.4 (thus the format is known also as S5E10).

12.4.4 Subband Encoding – JPEG HDR

JPEG HDR is an extension to the JPEG format for storing HDR images that is backward compatible with an ordinary 8-bit JPEG [5]. A JPEG HDR file contains a tone mapped version of a HDR image and a ratio (subband) image, which contains information needed to restore the HDR image from the tone

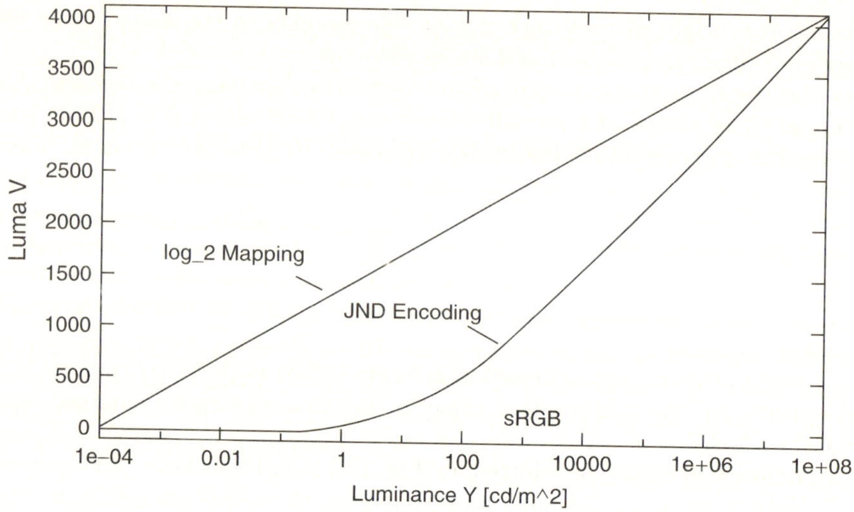

Fig. 12.11. Functions mapping physical luminance y to encoded luma values l. JND Encoding – perceptual encoding of luminance; sRGB – nonlinearity (gamma correction) used for the sRGB color space; logarithmic compression – logarithm of luminance, rescaled to 12-bit integer range. Note that encoding high luminance values using the sRGB nonlinearity *(dashed line)* would require a significantly larger number of bits than the perceptual encoding

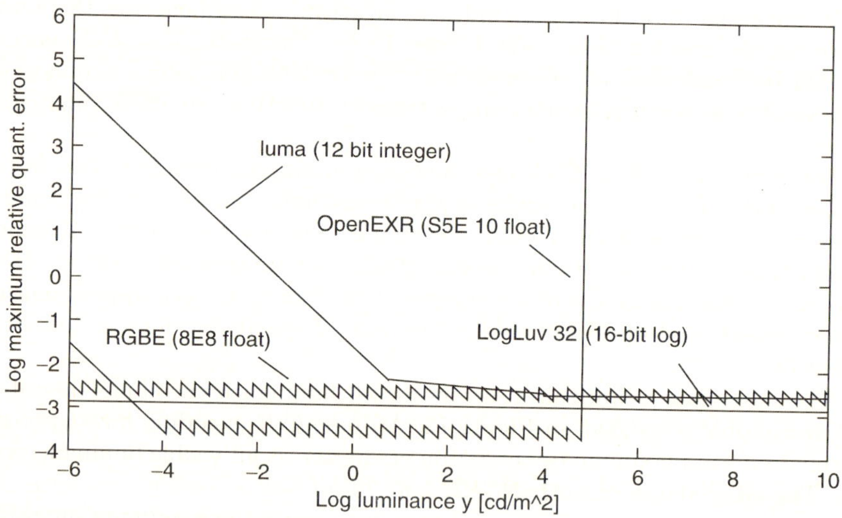

Fig. 12.12. Comparison of the maximum quantization errors for different luminance to luma encodings: JND encoding is given by 12.3; RGBE is an encoding used in the Radiance HDR format; 16-bit half is a 16-bit floating point format used in OpenEXR; 32-bit LogLuv is a logarithmic luminance encoding used in LogLuv TIFF format

but also reduces invisible noise in very dark scenes. Such noise reduction can significantly improve image or video compression.

12.7 Software for HDR Image and Video Processing

There is a growing number of applications that can read, write and modify HDR images and video. It is to be expected that most of the digital image processing software will support HDR data in the near feature. Currently however, (as of September 2005) the choice of applications that can operate on higher dynamic range is rather limited. Therefore, we list some free and commercial software that can take advantage of HDR:

- pfstools — a set of command line programs for reading, writing, manipulating and viewing HDR images and video frames. The software facilitates image file format conversion, resizing, cropping and rotating video frames or images. It is available as open source under GPL. URL: `http://pfstools.sourceforge.net/`
- OpenEXR — a library for handling OpenEXR files and a viewer for HDR images (exrdisplay). The library is available as open source. URL: `http://www.openexr.org/`
- Photosphere — an HDR Photoalbum software for Mac OS X, which can also create HDR images using multi-exposure technique. Available for free. URL: `http://www.anyhere.com/`
- Photogenics HDR — an image editing and paiting program, which can operate on 32-bit buffers. The program addresses similar applications as Adobe Photoshop. It is commercially available. URL: `http://www.idruna.com/photogenicshdr.html`
- HDR Shop 2 — can make HDR images from LDR photographs using multi exposures technique. Some image processing operations are included as well. It is commercially available. URL: `http://www.hdrshop.com/`

References

1. ICC. *Specification ICC.1:2004-10 (Profile version 4.2.0.0) Image technology colour management – Architecture, profile format, and data structure.* International Color Consortium, 2004
2. Greg Ward. Real pixels. *Graphics Gems II*, pp. 80–83, 1991
3. G. Ward Larson. LogLuv encoding for full-gamut, high-dynamic range images. *Journal of Graphics Tools*, 3(1):815–830, 1998
4. R. Bogart, F. Kainz, and D. Hess. OpenEXR image file format. In *ACM SIGGRAPH 2003, Sketches & Applications*, 2003
5. Greg Ward and Maryann Simmons. Subband encoding of high dynamic range imagery. In *APGV '04: Proceedings of the 1st Symposium on Applied perception in graphics and visualization*, pp. 83–90, New York, NY, USA, 2004. ACM Press

6. Rafał Mantiuk, Grzegorz Krawczyk, Karol Myszkowski, and Hans-Peter Seidel. Perception-motivated high dynamic range video encoding. *ACM Trans. Graph*, 23(3):733–741, 2004
7. William B. Thomspon, Peter Shirley, and James A. Ferwerda. A spatial post-processing algorithm for images of night scenes. *Journal of Graphics Tools*, 7(1):1–12, 2002
8. Rafał Mantiuk, Karol Myszkowski, and Hans-Peter Seidel. Lossy compression of high dynamic range images and video. In *Proc. of Human Vision and Electronic Imaging XI*, volume 6057 of *Proceedings of SPIE*, page 60570V, San Jose, USA, February 2006. SPIE
9. Charles A. Poynton. *A Technical Introduction to Digital Video*, Chapter 6, Gamma. John Wiley & Sons, 1996
10. M. Stokes, M. Anderson, S. Chandrasekar, and R. Motta. A standard default color space for the internet – sRGB. 1996
11. CIE. *An Analytical Model for Describing the Influence of Lighting Parameters Upon Visual Performance*, volume 1. Technical Foundations, CIE 19/2.1. International Organization for Standardization, 1981

13

HDR Applications in Computer Graphics

Michael Goesele and Karol Myszkowski

13.1 Introduction

One of the goals of computer graphics is to realistically render real objects and scenes taking their physical properties into account. Due to the high dynamic range (HDR) encountered in real scenes this requires performing most rendering steps with floating point precision. This was acknowledged early on and implemented in many software rendering systems. But the limited computational resources in hardware accelerated rendering systems often required developers to work with lower precision – typically 8 bit per color channel. With the advent of more powerful graphics cards, CPUs, and cluster rendering systems, an increasing number of systems incorporate now a full floating point rendering pipeline so that dealing with HDR data has become a standard requirement in computer graphics. One indication for this development is the now wide-spread use of HDR image file formats such as the OpenEXR format [1].

Without exact scene models (including materials, objects, lighting), realistic and physically correct rendering is, however, impossible. Graphics research is therefore also introducing methods to digitize various aspects of reality. Many of these methods are using camera systems as input devices and are therefore called *image-based methods*. One important aspect is hereby the ability to cope with the often huge dynamic range of the captured scene either using software or hardware system. In this chapter, we first discuss common approaches in the computer graphics area for capturing and calibrating HDR image data. We then describe several approaches for the digitization of real-world objects (Sect. 13.3) and for image-based lighting (Sect. 13.4). The chapter is concluded by a discussion of the requirements for HDR capture systems in computer graphics applications (Sect. 13.5).

13.2 Capturing HDR Image Data

One of the earliest examples of HDR imaging was introduced by Wyckoff and Feigenbaum [2] in the 1960s who invented an analog film with several emulsions of different sensitivity levels. This false color film had an extended dynamic range of about 10^8 which is sufficient to capture the vast majority of scenes encountered in practice without changing any exposure settings (aperture, exposure time) or adding filters to the optical path. An additional advantage is that brightness levels can be easily compared between different images provided that camera settings remain unchanged.

13.2.1 Multiexposure Techniques

As in the case of analog film, the dynamic range of traditional CCD imaging sensors often used in computer graphics applications is quite limited (typically in the order of 10^3–10^4). Several authors proposed therefore methods to extend the dynamic range of digital imaging systems by combining multiple images of the same scene that differ only in exposure time (*multiexposure techniques*). Madden [3] assumes linear response of a CCD imager and selects for each pixel an intensity value from the brightest nonsaturated image. This value is scaled according to the image's exposure and stored in the final HDR image.

A large number of methods has been proposed by several authors including Mann and Picard [4], Debevec and Malik [5], Mitsunaga and Nayar [6] and Robertson et al. [7] that estimate the *Opto Electronic Conversion Function* (OECF) (mostly called *response curve* in computer graphics) from a set of images with varying exposure. Given the response curve, input images can be linearized and scaled according to their exposure settings. The final HDR image is computed as a weighted sum of these images, where the weighting function reflects the certainty with which the value of an individual pixel in any of the input images is known (e.g., pixel values close to over- or underexposure are assigned a low certainty and consequently a low weight).

Although capturing multiple images of the same scene with different exposure settings has several disadvantages (e.g., long acquisition times due to the multiple exposures, difficulties for dynamic scenes), it is used in many applications. Systems using cameras with native HDR sensors, which can avoid many of these problems, are only slowly gaining momentum (see also the discussion of camera requirements for computer graphics applications in Sect. 13.5).

13.2.2 Photometric Calibration

For many applications, photometric calibration of the captured image data is crucial. Most multiexposure techniques yield images with linear floating point data per pixel in unknown units. These can be related to physical quantities by performing absolute calibration with a test target with known luminance (e.g.,

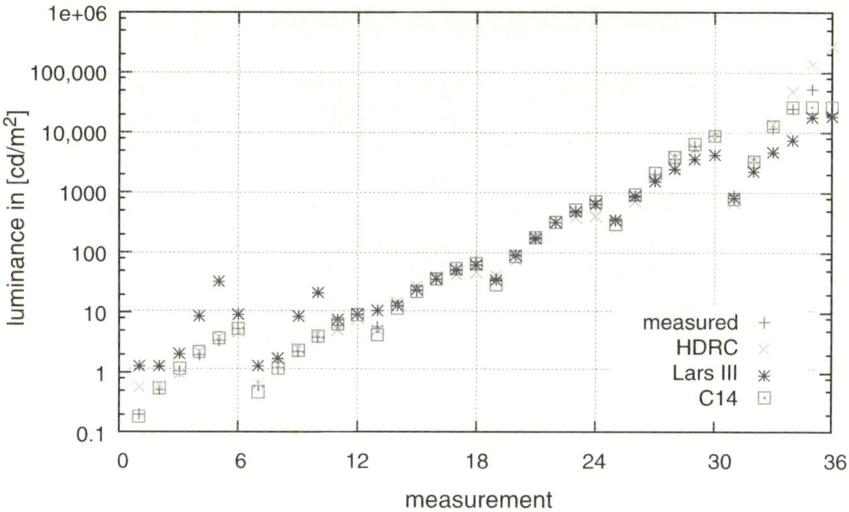

Fig. 13.1. Absolute calibration of several camera systems. Gray patches were illuminated with different intensities and their absolute luminance values were determined using a luminance meter. Absolute luminance values were then captured using a standard CCD camera (Jenoptik C14) with multiexposure technique and using two HDR cameras (IMS-Chips HDRC, SiliconVision Lars III)

determined by a light meter). Similarly, imagery from HDR camera systems needs to be linearized and calibrated to yield absolute luminance values.

To determine the accuracy of various HDR capture methods, Krawczyk et al. [8] used a series of test targets with known luminance values covering about 6 orders of magnitude. The luminance values were determined using a standard luminance meter. HDR images of these targets were acquired with multiexposure techniques using a standard CCD camera (Jenoptik C14). As shown in Fig. 13.1, the camera could capture all target luminance values with high accuracy apart from the two brightest samples. These samples were overexposed in all input images and no valid data could be recovered. They furthermore performed an absolute calibration of two HDR camera systems (IMS-Chips HDRC, SiliconVision Lars III) using two methods: an adapted version of a standard multiexposure calibration [7] and a simple fitting procedure based on the known linear or logarithmic camera behavior as given in the manufacturers' data sheets. Figure 13.1 shows also the results of capturing the same test targets with the calibrated HDR cameras. For valid samples within a defined luminance range, the accuracy of the HDR cameras calibrated with the latter approach is comparable to the multiexposure technique with an average error of about 6%. But the accuracy falls off for samples taken with the HDR cameras in very bright or very dark conditions, whereas the multiexposure technique yields data with constant accuracy as long as some of the input images contain valid data for a sample location.

takes into account soft shadows and interreflection effects. The outgoing radiance can be simply computed as the dot product between the incident lighting and radiance transfer vectors of SH coefficients. For Lambertian surfaces the lighting and transfer vectors composed of 25 SH coefficients for each sample point lead to good visual results for slowly changing and smooth lighting. For more general reflectance functions for which the incoming lighting directions are important, a matrix of spherical harmonic coefficients with the transfer vectors for each of those directions must be considered. In practice, matrices of 25×25 coefficients are commonly used [35]. Since the transfer vectors (matrices) are stored densely over the scene surfaces (usually for each mesh vertex) data compression is an important issue. It can be efficiently performed using standard tools such as principal component analysis (PCA) and clustering [36]. Recently, the limitation of low-frequency lighting has been lifted using the wavelet basis functions [37]. Using this approach proposed by Ng et al. both soft and sharp shadows can be rendered, but a very dense mesh is required to reconstruct the lighting function precisely.

PRT techniques are suitable for real-time VEM-based lighting of static scenes, for which the transfer vectors are precomputed at the preprocessing stage and the projection of environment map lighting into the SH basis is inexpensive and can be performed for each frame. Soft shadows, interreflections, and even more advanced shading effects such as subsurface scattering and translucency can be efficiently rendered using those techniques. It is easy to extend PRT techniques for dynamic scenes with prior knowledge of the animation by precomputing the radiance transfer data for each frame (or keyframe), which involves huge storage costs. However, efficient handling of fully dynamic scenes in interactive scenarios is still an open research issue.

Environment Maps Importance Sampling

For efficiency reasons many interactive applications rely on rendering based on graphics hardware. The best performance of such rendering in terms of the direct lighting and shadow computation can be achieved for directional light sources. It is thus desirable that captured environment maps are decomposed into a set of representative directional light samples. A number of techniques based on the concept of importance sampling have been recently proposed to perform such a decomposition for static lighting [38–40]. Unfortunately, their direct extension for VEM is usually not feasible due to at least one of the following major problems:

- Too high computational costs precluding VEM capturing and scene relighting at interactive speeds [38, 39]
- Lack of temporal coherence in the positioning of selected point light sources for even moderate and often local lighting changes in VEM [38,40]
- Lack of flexibility in adapting the number of light sources to the rendered frame complexity as might be required to maintain a constant frame rate [40]

The latter two problems, if not handled properly, can lead to annoying popping artifacts.

Havran et al. [41] have proposed an efficient importance sampling algorithm which leads to temporally coherent sets of light sources of progressively adjustable density. In their algorithm they treat the pixel luminance values in the EM as a discrete 2D probability density function (PDF) and they draw samples (light source directions) from this distribution following procedures established in the Monte Carlo literature. For this purpose they select an inverse transform method [42], which exhibits unique continuity and uniformity properties that are desirable for VEM applications. The method guarantees the bicontinuity property for any nonnegative PDF, which means that a small change in the input sample position over the unit square is always transformed into a small change in the resulting position of light source over the EM hemisphere. The uniformity property is important to achieve a good stratification of the resulting light source directions.

To reduce temporal flickering in the resulting frames Havran et al. choose the same set of initial samples over the unit square for each VEM frame, which are then processed using their inverse transform method. Since their emphasis is on interactive applications, they use a progressive sequence of samples in which adding new samples does not affect the position of samples already used for shading computations, while good sample stratification properties are always preserved. This is important for an adaptive selection of the number of light sources to keep a constant frame rate and for the progressive image quality refinement where the number of light sources can be gradually increased (this requires freezing the VEM frame). Havran et al. achieve the progressiveness of sampling sequence using quasi-Monte Carlo sequences such as the 2D Halton sequence. They perform Lloyd's relaxation over the initial sample positions in the unit square in a preprocessing step. This allows us to achieve a blue noise sample pattern on the hemisphere.

Since even local changes in the EM lead to global changes of the PDF, the direction of virtually all light sources may change from frame to frame, which causes unpleasant flickering in the rendered images. Havran et al. apply perception-inspired, low-pass FIR filtering to the trajectory of each light motion over the hemisphere and also to the power of the light sources.

13.4.2 A CAVE System for Interactive Global Illumination Modeling in Car Interior

Global illumination dramatically improves realistic appearance of rendered scenes, but is usually neglected in virtual reality (VR) systems due to its high costs [43]. In this section we present a VR system for realistic rendering of car interiors illuminated by VEMs that have been captured for various driving conditions and are visible through the car windows (refer to Fig. 13.5). The main goal of the system is to investigate the impact of quickly changing lighting conditions on the visibility of information displayed on LCD panels

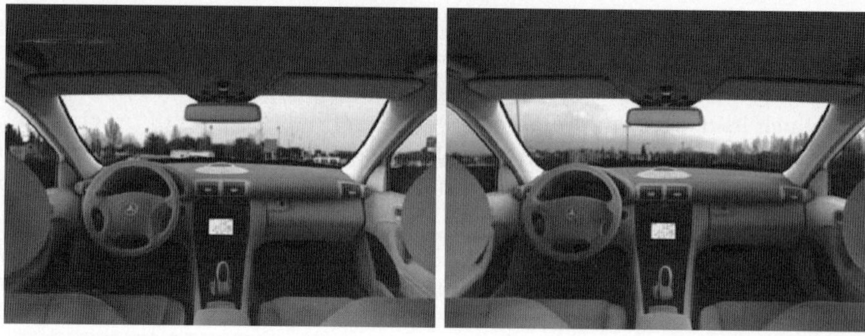

Fig. 13.5. Interactive rendering of car interior for two different lighting conditions

(commonly mounted on the dashboard in modern cars). This requires a global illumination solution responding interactively to lighting changes, which result from different car orientations in respect to distant lighting stored as dynamically changing VEM. This application scenario is similar to the simulation of free driving in an environment in which buildings, trees, and other occluders change the amount lighting penetrating the car interior. In such a scenario the response for lighting changes should be immediate for an arbitrary observer (virtual camera) position, but it can be safely assumed that the geometry of car interior is static, which greatly simplifies the choice of global illumination solution. To improve the immersion experience a CAVE system is used for displaying the car interior. A head tracking system monitors the current observer position, to properly model light reflection in the LCD panel.

The choice of global illumination solution in this application is strongly constrained by the hardware configuration of the CAVE system. Each of the five walls in the test system is powered by two consumer-level dual-processor PCs needed for the stereo projection effect (i.e., 5×2 frames must be simultaneously rendered). Each PC is equipped with a high-end graphic card. The design goal is to select such GI algorithms that fully exploit the computational power of the available CPUs and GPUs and can produce images of high (full screen) resolution at interactive rates. The choice of global illumination algorithm is also influenced by major characteristics of the car interior: static complex geometry and dominating Lambertian lighting interreflection component.

Taking into account all discussed system constraints, PRT techniques (refer to "Precomputed Radiance Transfer"), which require very costly preprocessing, but then enable real-time rendering that is particularly efficient for environments with predominantly diffuse reflectance properties.

The accuracy of light reflection modeling in the LCD panel is the most critical requirement in this VR system. Since the LCD panel reflectance involves also glossy component, the low-frequency lighting assumption that is imposed by view-independent PRT techniques might not hold so well. To

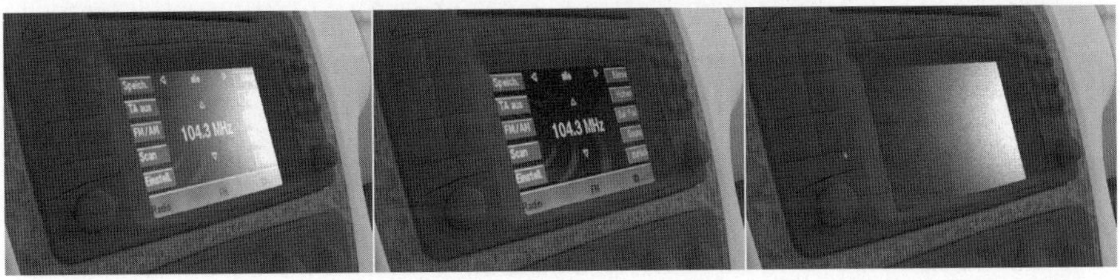

Fig. 13.6. The LCD panel appearance as a result of the global illumination computation for VEM lighting (Left: full global illumination, Center: display emitted light only, Right: reflected light)

improve the spatial resolution of lighting details for a given camera view *final gathering* is used, in which the results obtained using the PRT techniques are stochastically integrated for selected sample points. To reduce the variance of such integration the reflectance function of LCD panel is used as an important criterion for choosing sample directions. Figure 13.6 shows the appearance of a LCD panel under global illumination conditions for dynamic VEM lighting as displayed on a HDR monitor. Since the dynamic range of the HDR monitor is significantly higher than the one of a typical LCD panel that is mounted in the car cockpit, the visibility of information displayed for the driver can be tested for many external lighting conditions. Through the calibration of the HDR display, real-world luminance values can be reproduced for the LCD panel taking into account both the panel emissivity as well as reflected lighting resulting from the global illumination simulation.

Due to separate treatment of the car interior and LCD panel, a hybrid global illumination approach is implemented in which the PRT lighting computation and rendering are performed on GPUs for all five CAVE walls in stereo. The final gathering computation also uses the results of PRT lighting, but it is performed in parallel on all idle CPUs in the system. This way the final gathering reputed as a computationally heavy off-line technique can be performed with interactive performance in the VR system.

13.4.3 Interactive Lighting in Mixed Reality Applications

Havran et al. [41] have presented a complete pipeline from the HDR VEM acquisition to rendering at interactive speeds. The main technical problem in this work is efficient processing of the acquired HDR video, which leads to a good quality rendering of dynamic environments using graphics hardware.

For the HDR VEM acquisition Havran et al. use a photometrically calibrated HDRC VGAx (IMS CHIPS) camera with a fish-eye lens. They decompose each captured frame of VEM into a set of directional light sources, which are well suited for the shadow computation and shading using graphics hardware (refer to "Environment Maps Importance Sampling"). The light sources after temporal filtering are used in the pixel shader of a GPU. The main computational bottleneck is the shadow map computation for hundreds

Fig. 13.7. *Left*: the model with strong specular reflectance properties composed of 16,200 triangles rendered with 72 shadow maps at 5.3 Hz. On the *left top* is an environment map captured in real time using a HDR camera with fish-eye lens. The light sources are marked by *green points*. On the *left bottom* the same environment map is shown in polar projection. *Right*: the same model illuminated by outdoor lighting

of light sources. They propose an algorithm that for a given extent of the normal vector directions of a shaded pixel efficiently eliminates most of the invisible light sources. Optionally, they cluster lights with similar directions corresponding to the regions of concentrated energy in the VEM, e.g., around the sun position, to reduce the shadow computation cost even further.

The system presented by Havran et al. lifts common limitations of existing rendering techniques, which cannot handle at interactive speeds HDR image-based lighting captured in dynamic real-world conditions along with complex shadows, fully dynamic geometry, and arbitrary reflectance models evaluated on a GPU (refer to Fig. 13.7 in the color plate). The system runs on a PC with a 3 GHz Pentium4 processor and NVidia GeForce 6800GT graphics card. The performance of 22 fps for the VEM capturing and directional light generation with filtering is achieved. The asynchronous rendering performance drops to 7 fps when 72 light sources and shadows at the image resolution of 320×240 pixels are considered.

The system developed by Havran et al. has many potential applications in mixed reality and virtual studio systems, in which real and synthetic objects are illuminated by consistent lighting at interactive frame rates (refer to Fig. 13.8 in the color plate). The synthetic entities are expected to blend seamlessly with the real-world surroundings, objects, and persons. So far dynamic lighting in such systems is performed by scripting rigidly set matrices of light sources whose changes are synchronized with the virtual set rendering. Such systems are expensive and require significant human assistance, which reduces their applicability for smaller broadcasters. The presented system can support arbitrary changes of lighting in virtual studio in a fully automatic way.

13.5 Requirements for HDR Camera Systems

The earlier sections showed some applications of HDR camera systems in computer graphics. In the following, we summarize some of the requirements

Fig. 13.8. Comparison of the fidelity in the shadow and lighting reconstruction for the real-world and synthetic angel statuette illuminated by dynamic lighting. Real-world lighting is captured by the HDR video camera located in the front of the round table with an angel statuette placed atop (the *right image side*). The captured lighting is used to illuminate the synthetic model of the angel statuette shown in the display (the *left image side*)

for HDR camera systems that are used in such applications. Their importance depends naturally on the application at hand and such a list will therefore never be complete.

High dynamic range. As shown previously, the limited dynamic range of many camera systems is a key problem in computer graphics. Having sufficient dynamic range available (a dynamic range of 10^6 and above seems to be sufficient for most current applications) is therefore mandatory.

High sampling depth. Quantization artifacts can cause problems in numerical computations or lead to visible banding artifacts. The individual pixel values should be digitized with highest possible sampling depth. Note that the meaningful sampling depth will also be limited by factors such as noise or optical blurring due to the point spread function of the imaging system.

High image resolution. Spatial detail in captured images is often more important than overall structure (e.g., for captured textures). Graphics application require therefore often higher image resolutions than applications from other fields.

Good image quality. The quality of the input images plays a huge role for the achievable results – especially if the results are not interpreted but displayed directly (e.g., in the form of a texture). Standard requirements such as low noise or good color calibration are therefore also important for the case of HDR images.

33. Ulf Assarsson and Tomas Akenine-Möller. A geometry-based soft shadow volume algorithm using graphics hardware. *ACM Transactions on Graphics*, 22(3):511–520, 2003
34. Ravi Ramamoorthi and Pat Hanrahan. An efficient representation for irradiance environment maps. In *Proceedings of ACM SIGGRAPH 2001*, Computer Graphics Proceedings, Annual Conference Series, pp. 497–500, August 2001
35. Peter-Pike Sloan, Jan Kautz, and John Snyder. Precomputed radiance transfer for real-time rendering in dynamic, low-frequency lighting environments. *ACM Transactions on Graphics*, 21(3):527–536, 2002
36. Peter-Pike Sloan, Jesse Hall, John Hart, and John Snyder. Clustered principal components for precomputed radiance transfer. *ACM Transactions on Graphics*, 22(3):382–391, July 2003
37. Ren Ng, Ravi Ramamoorthi, and Pat Hanrahan. Triple product wavelet integrals for all-frequency relighting. *ACM Transactions on Graphics*, 23(3):477–487, 2004
38. Sameer Agarwal, Ravi Ramamoorthi, Serge Belongie, and Henrik Wann Jensen. Structured importance sampling of environment maps. *ACM Transactions on Graphics*, 22(3):605–612, 2003
39. Thomas Kollig and Alexander Keller. Efficient illumination by high dynamic range images. In *Eurographics Symposium on Rendering: 14th Eurographics Workshop on Rendering*, pp. 45–51, 2003
40. Victor Ostromoukhov, Charles Donohue, and Pierre-Marc Jodoin. Fast hierarchical importance sampling with blue noise properties. *ACM Transactions on Graphics*, 23(3):488–495, 2004
41. Vlastimil Havran, Miloslaw Smyk, Grzegorz Krawczyk, Karol Myszkowski, and Hans-Peter Seidel. Interactive system for dynamic scene lighting using captured video environment maps. In *16th Eurographics Symposium on Rendering*, pp. 31–42, 2005
42. Vlastimil Havran, Kirill Dmitriev, and Hans-Peter Seidel. Goniometric Diagram Mapping for Hemisphere. Short Presentations (Eurographics 2003), 2003
43. Kirill Dmitriev, Thomas Annen, Grzegorz Krawczyk, Karol Myszkowski, and Hans-Peter Seidel. A CAVE system for interactive modeling of global illumination in car interior. In Rynson Lau and George Baciu, editors, *ACM Symposium on Virtual Reality Software and Technology (VRST 2004)*, pp. 137–145, Hong Kong, 2004

14

High-Dynamic Range Displays

Helge Seetzen

We see more than we can process. This fact of our visual system applies equally to the HDR imaging pipeline. Digital cameras can capture more than displays can show. Likewise, most rendering techniques produce results that overwhelm the capabilities of conventional displays. Overall, there are few output devices today that can rival the dynamic range of comparable input devices. This chapter describes techniques to overcome this limitation as well as an implementation example in the form of the BrightSide Technologies DR37 HD HDR Display (Fig. 14.1).

14.1 HDR Display Requirements

Any effective HDR Display technology needs to deliver not only high dynamic range output but also meet key industry and infrastructure requirements. Specifically, the HDR Display technology should be:

- Able to present realistic luminance distributions
- Compatible with existing imaging infrastructure
- Reasonably cost effective

The latter two requirements are the easiest to define. Any improved display technology needs to function within today's imaging pipeline. Ideally it should also provide some benefit over existing solution even with conventional input. An example of a novel approach that did not meet this requirement is high definition (HD) television. These displays, while offering compelling benefits with the right content, deliver little improvement when showing much more common conventional (regular definition) footage. As a result the introduction of HD displays has been very slow and only the stamina of very large consumer electronics companies allows HD Displays to compete in the market today.

The cost effectiveness requirement is also quite self-explanatory. While higher performance displays can command some price premium, they need to be in the same league with conventional solutions to leave the niche application space.

The first requirement is intuitively the most obvious. Clearly, HDR Displays should be able to present images with realistic luminance distributions. A matching technical description for this objective requires a good definition of the human visual system (HVS) and the translation thereof into display performance. The latter is particularly challenging because of variable terminology used in the display industry.

Natural scenes offer a vast luminance range over 14 orders of magnitude from starlight to sunlight. Our visual system has evolved to operate in this environment by using adaptation mechanism. For the purpose of HDR Displays the ambient environment is usually a darkened room (e.g. home theatre) or an office setting. Comfortable peak luminance levels for such an environment are 3,000–6,000 $\mathrm{cd\,m^{-2}}$ depending on the size of the feature [1]. A good HDR Display should reach this level to maximize the use of the viewer's capabilities.

With the peak luminance of the display set in this fashion, the next step is to define the black luminance and contrast of the display. The use of these terms varies greatly between psychophysics and the display industry. Among display vendors the contrast specification is often provided with qualifiers such as "on/off contrast" (ratio between a completely black and a completely white screen), "frame sequential contrast" (contrast achievable over multiple frames) or "ANSI contrast" (ratio between black and white portions of a checkerboard pattern). Each metric yields different numbers. Most display technologies have fairly low ANSI contrast due to light leakage from bright into dark regions of the checkerboard. On/off contrast is usually the highest value for devices that can increase or reduce the light output if the entire screen in black (though of course it is arguably also the least relevant metric for actual video content since the presence of some white pixel in most normal video frames prevents the display from ever reaching this exaggerated black level). Rather than agonizing over the correct choice, it is simpler to conclude that there is nothing wrong with having a black level near or at zero. This sets the ideal luminance range of HDR Displays at 3,000–6,000 $\mathrm{cd\,m^{-2}}$ to 0 $\mathrm{cd\,m^{-2}}$ for applications where visually pleasant images are desired. Obviously this numbers will change if accurate luminance reproduction of very high brightness features is desired in applications such as scientific simulation.

The final aspect of the technical requirement is the amplitude resolution of the display. Like contrast, this is a difficult characteristic that depends highly on the HVS characteristics. The HVS has a non-linear response and as a result most displays feature at least some form of gamma response. Fortunately, psychophysics provides a good description of the contrast sensitivity threshold as a function of luminance. The amplitude resolution and response curve of a good HDR Display then should be such that the digital step size at any luminance level is lower than the perceivable threshold at that luminance.

14.2 HDR Display Design

There are in principle two avenues to design HDR Displays. Either each pixel of the display can be modulated accurately over a very wide range or the output of a pixel is actually the serial combination of two or more modulators. In the later case the final dynamic range of the display is given by the product of the ranges of each of the sub-modulators.

Technologies in the first category include laser projection and organic light emitting diodes (OLED). These devices are at least in principle capable of modulating each pixel from zero to some high luminance value. In reality their capabilities are more restricted. In both cases the largest restriction is the limited peak luminance of the devices. Laser projection requires high power laser diodes which are currently prohibitively expensive, especially for green and blue lasers. The materials for OLED devices are available but suffer from low efficiencies and significant degradation over time. In both cases these limitations have prevented the technology from reaching the luminance level of today's displays, much less the level required for HDR imaging.

Another serious limitation for direct approaches is the availability of high amplitude resolution drive circuitry. A wide range of luminance is only useful if a reasonable number of luminance levels are distinctly addressable within this range. Useful HDR Displays require an amplitude resolution or bit depth ranging from 12 to 16 bit depending on the application. High bit depth drivers of this kind are currently expensive and challenging to implement.

Nevertheless, both approaches show promise as materials and drivers continue to improve. Proponents of these concepts estimate 10–15 years for laser projection to become available for consumer applications and 3–5 years for OLED displays to reach the luminance level of today's conventional displays. An improvement beyond today's level will likely take significantly more time and might not be possible at all given the physical limit on OLED material efficiency.

The second approach offers a path to reach HDR image quality with existing devices. Combining two LDR devices removes the need for high bit depth drivers though high bit depth image processing logic is still required. The following describes implementations of dual-modulation displays and projectors, the associated image processing algorithms and the performance that can be obtained in this fashion.

The alternative are displays that use multiple lower bit depth modulators in series. Such dual-modulation HDR Displays can be built by combining conventional transmissive displays with a spatially modulated light source. Transmissive displays such as liquid crystal displays (LCD) are passive devices that allow a portion of incident light to pass through each pixel. The transmission can be modulated over a reasonable range between 200:1 and 500:1. Standard LCD are combined with a passive light source such as cold cathode fluorescent tubes (CCFL) which provide a uniform and constant light field at the back of the LCD. A spatially modulated backlight can greatly increase the dynamic

Fig. 14.1. BrightSide Technologies DR 37 HD HDR Display

Fig. 14.2. HDR display image processing example. From top left: the LED backlight shows a low resolution approximation of the luminance distribution of the image; the actual light distribution from the LED backlight is blurred by the optical package; the LCD image compensates for the low resolution of the LED backlight; the final image is the product of the LED light distribution and the LCD image

range of such a display. Such a backlight could be a secondary LCD (or other type of transmissive modulator) in front of a high brightness light source or an array of individually controlled light sources such as light emitting diodes (LED).

High quality LCD have a very low transmission efficiency even in the white state (approximately 2–5%) and combining two transmissive modulators is therefore very inefficient. The LED approach is therefore by far the most energy efficient because the efficiency loss of the second LCD is avoided. Moreover, in the LED approach light is only created in the bright regions of the image. The average power consumption of LED based HDR Displays is therefore lower than that of conventional LCD where the backlight is constantly on. Normal TV signal has an average intensity level of approximately 22% such that the LED approach is on average 3–5 times more power efficient than a conventional LCD display of the same brightness. The efficiency gain compared to the dual modulation LCD approach is even more significant.

An optimized HDR Display design has consequently three major components: a backlight with a low resolution array of LED and appropriate optical structures, a high resolution LCD panel, and an image processing algorithm to drive both modulators (Fig. 14.2). Each component uses standard electronic and optical elements so that the implementation and design cost are low.

14.2.1 LED Backlight

The LED backlight's primary function is to provide a smooth light distribution. This can be achieved with a point spread function (PSF) of the light emitted by a single LED that overlaps significantly with the PSF of neighboring LED. This blurs the light field created by the LED backlight sufficiently to remove any high spatial frequency information which could interfere with the high resolution pixel structure of the LCD. It also ensures that the backlight produces an approximately uniform light field if many LED are driven at the same level. The HDR Display can therefore operate like a conventional LCD display with a constant (and uniform) backlight if HDR image quality is not desired for a particular application.

Because the LED light distribution has radial symmetry it is desirable to arrange the LED in a hexagonal rather than rectangular grid. This arrangement will not interfere with the rectangular structure of the LCD pixel layout because of the vast resolution difference and blur of the LED light field. A hexagonal LED spacing of 10–30 mm is acceptable for monitor or TV applications provided the PSF is arranged such that it overlaps substantially with that of direct neighbors and falls off towards zero when it reaches secondary neighbors.

A reflective cavity between the LED and LCD will create such an optimized PSF. Conventional LCD use optical films such as brightness enhancement film [2], reflective polarizer film [2] and various diffuser sheets to shape the light distribution leaving the displays. These films have different functions

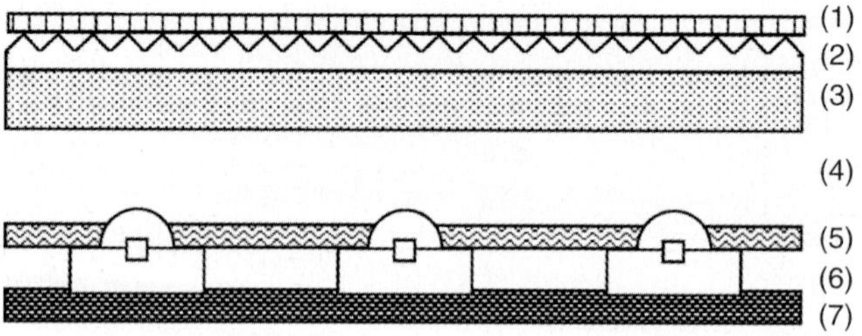

Fig. 14.3. Optical package of the HDR Display including LCD (1), micro-structured films such as BEF or DBEF (2), diffuser (3), cavity (4), reflective film (5), LED (6) and circuit board with appropriate thermal management solution (7)

but share the characteristic that they reflect some portion of the incoming light back towards the backlight. In the HDR display a significant portion of the light coming from the LED will therefore be reflected by the film layers back towards the LED. A reflective cavity can be created by placing a layer of reflective material onto the LED array. This causes the light from the LCD to bounce between the LCD (and film layers) and the back reflector. Some light leaves the LCD with each bounce but the reflected light also moves laterally so that the final PSF is very similar to a Gaussian or combination of Gaussian distributions. Fig. 14.3 shows a schematic diagram of this optical package.

The LED backlight also needs to be controllable over the entire range of each LED with some reasonable amplitude resolution. Because high spatial frequency and color modulation occurs in the LCD, the amplitude resolution of the LED control does not have to be very high. An 8-bit control component with 255 distinct steps is more than sufficient. Lower resolution control is also possible but usually it is cheapest to use mass produced 8-bit devices. Some of the excess amplitude can be used, for example, for calibration of the LED over lifetime or temperature.

14.2.2 LCD Panel

The LCD panel used for the HDR Display is a can be a standard panel. The LCD is responsible for high spatial frequency information, local contrast and color. Therefore the benefits of a better LCD will usually directly improve the HDR Display as well.

14.2.3 Image Processing Algorithm

Image processing for the HDR Display would be easy if the LED were small enough to match the LCD pixel in size. This is not the case and the image processing algorithm therefore needs to compensate for the low resolution of the LED backlight by adjusting the LCD image. This is possible because

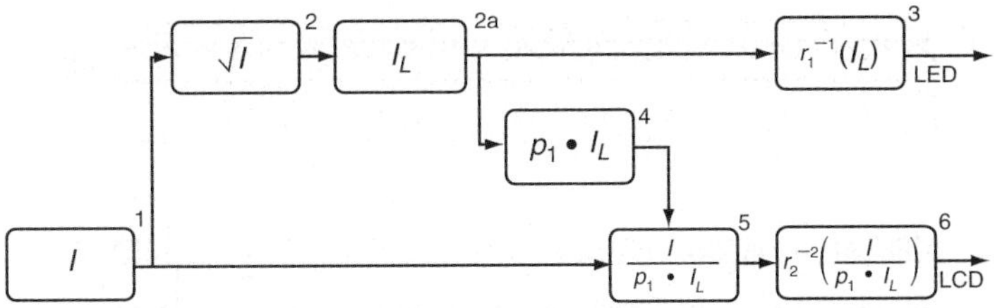

Fig. 14.4. HDR Display image processing algorithm

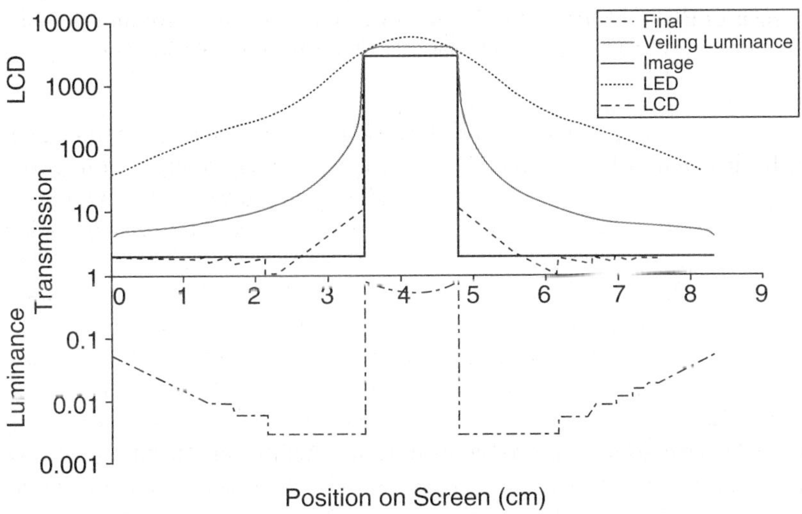

Fig. 14.5. High contrast feature showing undesirable LED halo masked by veiling luminance halo

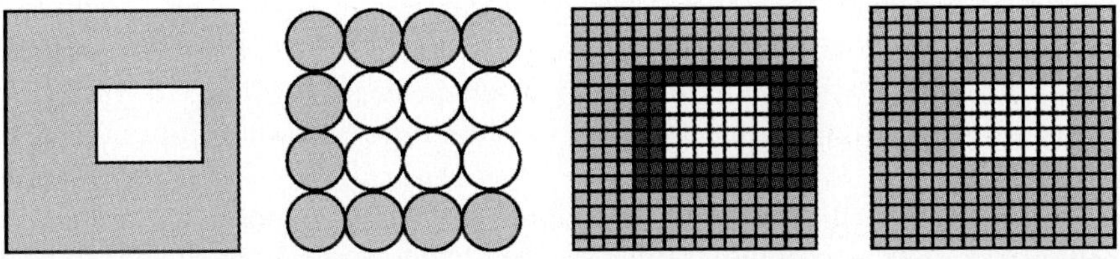

Fig. 14.6. Schematic HDR Display image processing steps: Input image, LED values, LCD image with correction, final output

the light field created by the LED backlight is effectively multiplied by the transmission pattern on the LCD.

The algorithm shown in Fig. 14.4 separates the incoming image (1) into LED (3) and LCD (6) information such that the optical multiplication of the two modulators returns the luminance levels defined by the input.

The first step in this process is to reduce the resolution of the incoming image to approximately that of the LED array. Since the LED array usually has a hexagonal layout this will involve some asymmetric down-sampling. Different methods can be used for this process depending on the capabilities of the processor. Once the image has been reduced to the LED layout, the algorithm decides what portion of the total image will be presented by the LED backlight. The final image is the product of the LED and LCD image so that an approximately equal distribution can be achieved by taking the square root (2) of the reduced resolution image (2a). Varying the power of the root will change this relationship. Likewise, non-linear response of the LCD and other factors such as thermal management characteristics will influence this decision.

The down-sampling and distribution steps are loosely constrained because many choices for the LED values will yield acceptable results. The LCD image will compensate for most of the variation in the LED backlight. The same reason allows generous application of common image enhancement techniques to the LED image such as sharpening filters and other techniques designed to optimize the energy efficiency of the display.

The next step is to calculate the anticipated light field (4) generated by the LED by summing up the contributions of each LED to the total light field. The PSF of each LED is adjusted by the drive value of the LED, the response function of the LED, and any other necessary factor such as thermal or lifetime calibration. It is important that the PSF used in this process accurately reflects the effective light distribution of one LED, including any interaction with the optics and films used in the package. The result of this summation is a simulation of the light field of the LED backlight for this input image.

Finally, the LCD image can be generated by dividing the input image by the light field simulation (5). If the LED backlight uses LED with different spectral components then these need to be simulated as well and the division can occur at the color channel level. For a white LED backlight the normally RGB input image will be divided by a monochrome light field. In both cases the result of the division are adjusted by the inverse response function of the LCD.

At this point the image processing algorithm is finished and the HDR Display can show a complete frame. The LED will be driven to the appropriate intensity and the LED backlight will create a luminance distribution that is similar to the light field simulation. The LCD will modulate this luminance distribution such that the final output will match the input image (Fig. 14.6).

In principle the algorithm can accurately reproduce any input image within the range of the product of the dynamic ranges of the LED and LCD elements. In actual devices there are limits to this approach because of the discreet step size of the LCD and LED, rounding errors in the simulation and a host of other factors. Specifically, if both LED and LCD are 8-bit devices with 255 discreet steps each then the total modulated range of 0–65025 but several steps inside that range are not achievable. For example, levels between 64,770($= 255 \times 254$)

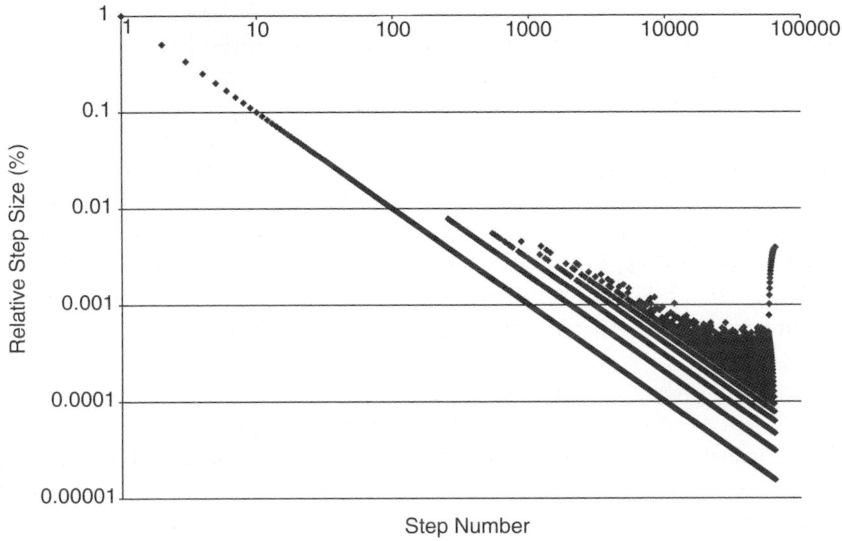

Fig. 14.7. Relative step size of dual-8-bit HDR display

and 65,025($= 255 \times 255$) are unreachable. As a result there are luminance levels that the HDR Display cannot reproduce accurately. Fortunately, the magnitude of these gaps in the luminance range of the HDR Display is very small at the low end of the range and increases towards the high end. The relative gap size is therefore very low everywhere along the range. Figure 14.7 shows this effect for a display with an 8-bit linear LED and an 8-bit LCD with a gamma of 2.2 as response curve. With the exception of the near black range the relative step size of the HDR Display is below 1%.

A further limitation of the HDR Display occurs at very high contrast boundaries. Such boundaries require the LED directly under it to be very bright, leaving only the LCD to adjust the image. The light field from the LED will be approximately constant in the region very close to the boundary so that the highest local contrast is the dynamic range of the LCD. As a result the bright side of the boundary can be displayed accurately but the dark side will be slightly grey near the boundary. Further away from the boundary full black is achievable because a different LED contributes to this area and it can be turned off. The HDR Display is therefore capable of modulation over the dynamic range of the product of the LED and LCD ranges globally but only that of the LCD locally.

Fortunately, this limitation is not relevant in conventional imaging. While our visual system is extremely good at dealing with high contrast images, it is limited in its capability in small areas. This is not so much an issue of our visual process but a result of optics. The optical system of our eye, including the lens and intraocular fluid, is imperfect. Specifically, each element scatters some of the light that passes through it. Our eye therefore has a well

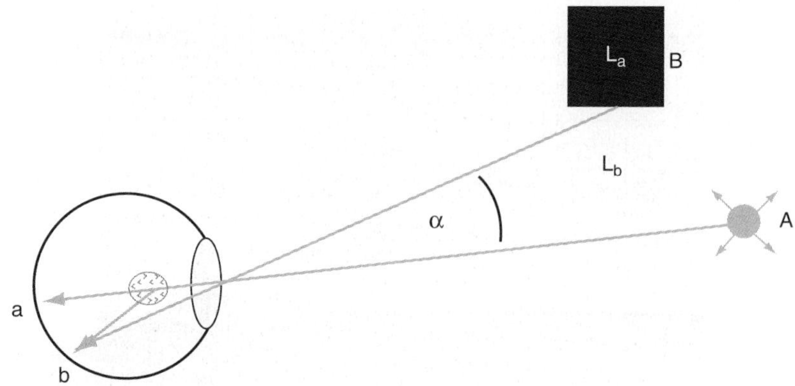

Fig. 14.8. Veiling luminance effect

established PSF for incoming light hitting the retina. A common model of this PSF is shown in (14.1). The PSF has two parts: a delta function term corresponding to the portion of incoming light that is not scattered, and a second term describing the scattered portion which is distributed as a function of the angle between the light source and the position on the retina. This is effect is often called veiling luminance or veiling glare.

Veiling luminance point spread function

$$P(\alpha) = \eta\delta(\alpha) + \frac{c}{f(\alpha)}. \tag{14.1}$$

Figure 14.8 shows a schematic outline of this effect. The scene is a high contrast boundary with bright region A producing luminance L_{a} and dark region B with luminance Lb. Scatter in the eye sends some of the light from A onto spot b on the retina where region B is imaged. The actual perceived luminance at b is therefore given by the contribution of L_{b} and the veiling luminance portion of L_{a} for the angle between A and B. This increase in the perceived luminance at b in turn lowers the perceived contrast between points A and B in the scene. Veiling luminance therefore puts a limit on the perceivable contrast in a small region. The LED spacing suggested earlier ensures that the halo caused by the large size of the LED will always be smaller than the veiling luminance halo caused by the bright portion of the high contrast boundary.

Figure 14.5 shows a cross section example of a small bright feature on black background. Notice the increased light level near the boundary resulting from the LCD's inability to compensate for the excess light from the LED. Also shown is the veiling luminance caused by the bright feature for an average viewer at a viewing distance of 20 cm. The veiling luminance effect is significantly stronger than the display imperfection. In other words, the HDR Display has a defect at high contrast boundaries but the HVS has a similar and larger defect at those points as well which results in the display defect becoming invisible.

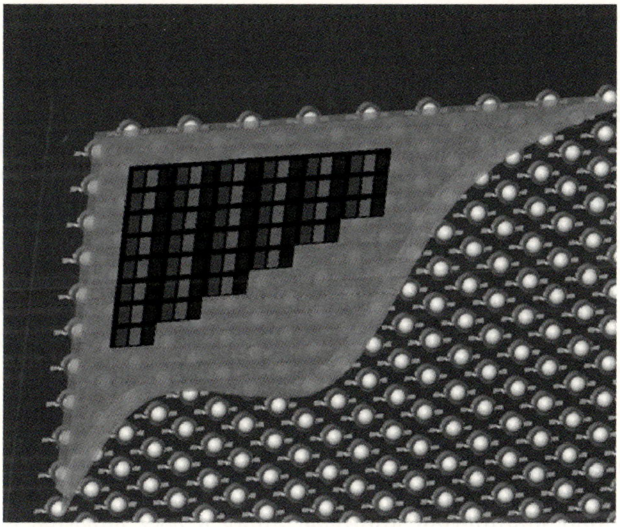

Fig. 14.9. HDR Display layout including LED backlight, optical package and LCD layer

14.3 HDR Display Performance

The dual-modulation HDR Display approach (Fig. 14.9) offers a wide range of improvements over conventional displays. First and foremost are of course the increases in dynamic range and amplitude resolution. But other benefits can be obtained as well. For example, the use of different color LED (or integrated RGB LED packages) can significantly improve the color gamut of the display by providing more saturated primaries. The LED have a very fast response time so that flashing them can increase the effective response time of the LCD (the LCD is a passive device so the effective light emissive period is defined only by the LED flash time). There are many other options that make use of the unique architecture of the HDR Display to enhance image quality or system efficiency.

BrightSide Technologies has developed a 37″ high definition HDR Display reference design that incorporates many of these benefits (Fig. 14.1). This unit offers a peak luminance of over 3,000 cd m^{-2} (approximately 5–10 times higher than any other TV on the market), a black level that is effectively zero when the LED switch off. The sequential contrast of such a device is infinite since the black level is zero. A more useful measure is the ANSI contrast which is approximately 25,000:1 on a 9-point ANSI checkerboard target. The non-zero ANSI black level is the result of small light contributions from LED below the bright regions towards the dark regions.

15

Appendix

15.1 Symbols

σ	the standard deviation representing the semiedge width of the detected edge;	
ξ_1	the inflection point coefficient that determines the brightness I_{turn} on the edge;	
ξ_2	the turning point coefficient that determines the background brightness I_{bgrd}	
η_0	the brightness factor that represents the minimal proportion between the brightness of the earliest possible edge position I_0 and the background I_{bgrd}	
$\bar{x}$, $\bar{y}$, $\bar{z}$	Color Matching Functions	
α	temperature gradient of the offset	
β	temperature gradient of the gain	
γ	temperature gradient of the illumination at the 3 dB-point	
$\sigma(\lambda)$	Sensitivity	
κ_{R}, κ_{G}, κ_{B}	R, G, B sensitivity	DN Dec^{-1}
κ_{R}, κ_{G}, κ_{B}	R, G, B sensitivity	DN Dec^{-1}
α_{R}, α_{G}, α_{B}	Color enhancement parameters	
$\bar{\text{Y}}$	Adapting luminance (background luminance in simple stimuli)	
a	offset	(DN)
a	Inverse contrast sensitivity	percent
a^*, b^*	Color vectors in CIELab	DN

a_0	Offset at 0° C	(DN)
a_{ph}	Physical offset	(V)
b	gain	
B	Brightness	
b_0	Gain at 0° C	
b_{ph}	Physical gain	
b_{R}, b_{G}, b_{B}	R, G, B offset	DN
c	Illumination at the 3 dB-point	(lx)
$C^*{}_{\mathrm{R}}$	Chroma	Dig.
c_0	Illumination at the 3 dB-point at 0° C	(DN)
C_{R}, C_{G}, C_{B}	Colorfulness red, green, blue output	Dig.
f_{R}, f_{G}, f_{B}	R, G, B OECF's	
I	Intesity	
I_{D}	Drain current	(A)
I_{Dark}	Dark current	(A)
I_{max}	Maximum current in weak inversion	(A)
I_{Photo}	Photo current	(A)
k	Normalization constant	
k	Boltzmann Constant	
k_{a}, k_{b}	CIELab Paramters	
l	Luma	
L	Brightness	
$l(y)$	Luminance to luma conversion function	
$L^*\ (x,\ y)$	Lightness	Dig.
L^*_{W}	Lightness white	Dig.
$L^{\rightarrow}(x_{\mathrm{o}}, \omega_o)$	Radiance leaving the surface at point x_{o} in direction ω_o	
lg	Logarithm base 2	
ln	Logarithm base e	
log	Logarithm base 10	
L_{W}	Brightness white	
n	subthreshold slope factor	
N	RMS noise	V, e, DN
NDFP	Dark fixed-pattern noise	V, e, DN
NDS	Dark shot noise	As, e, DN
NPS	Photosignal shot noise	As, e, DN
NQ	Quantisation noise	
NR	Reset noise	V, e, DN
NSFP	Photosignal fixed-pattern noise	V, e, DN
p	Central pixel	
q	Electron charge	As
Q_{D}	Dark charge	As, e, DN
Q_{S}	Signal charge	As, e, DN

$R(Y)$	Neural response to luminance Y	
R, G, B	Densitometric Tristimuli	DN
$\mathrm{Rd}(x_i, x_o)$	Diffuse reflectance	
$\mathrm{RF}(\bar{\mathrm{Y}})$	Highest resolvable spatial frequency for adapting luminance	
S	Signal	V, e, DN
$s(\lambda)$	Sensitivity	
$S(x_i, \omega_i; x_0, \omega_0)$	Bidirectional scattering-surface reflectance distribution function (BSSRDF)	
S_R, S_R^*	Saturation red	Dig.
T	Temperature	$^\circ$C, K
u'	first component of the CIE 1976 Uniform Chromacity Scales	
v'	Second component of the CIE 1976 Uniform Chromacity Scales	
$V(x, y)$	Spatially variant local adaptation value	
V_{GS}	gate-to-source voltage	(V)
V_R	Output voltage red	DN
V_t	Thermal voltage (0.025 V)	V
V_T	Threshold voltage (0.7 V)	
V_{thresh}	Threshold voltage	(V)
x	Illumination	(lx)
x	Chromaticity coordinate red	Dig.
x	Array coordinate	cm
x	Stimulus	
X	Colorimetric tristimulus value red	Dig.
x, y	Image coordinates	
$\bar{x}$, $\bar{y}$, $\bar{z}$	Color matching functions	
x_{bright}	Illumination at bright illumination (e.g. 1,000 lx)	(lx)
x_{dark}	Illumination at low illumination (e.g. 0.01 lx)	(lx)
x_i	Point at which light enters into the surface	
x_L, x_M, x_S	Colorfulness red, green, blue input	
x_{medium}	Illumination at medium illumination (e.g. 10 lx)	(lx)
x_o	point at which light leave the surface	
X_W, Y_W, Z_W	Tristimulus values white limit	

y	Digital output	(DN)
Y	Stimulus luminance	
y	Luminance	
y	Sensor output	V, DN
y	Chromaticity coordinate green	Dig.
Y	Colorimetric tristimulus value green	Dig.
$y(l)$	Luma to luminance conversion function	
y_{bright}	Digital output at bright illumination (e.g. 1,000 lx)	(DN)
y_{dark}	Digital output at low illumination (e.g. 0.01 lx)	(DN)
y_{medium}	Digital output at medium illumination (e.g. 10 lx)	
y_{raw}	Uncorrected, digital output	(DN)
Z	Chromaticity coordinate blue	Dig.
Z	Colorimetric tristimulus value blue	Dig.
$\phi^{\leftarrow}(x_{\mathrm{i}}, \omega_{\mathrm{i}})$	Incident flux entering into the surface at point x_{i} from direction ω_{i}	
ω_{i}	Incoming light direction	
ω_{o}	Outgoing (reflected) light direction	

List of symbols(E)			
		Units	Typ. range
AVDD	Analog supply voltage	V	3.3
AVSS	Analog ground	V	0
DVDD	Digital supply voltage	V	2.5
DVSS	Digital ground	V	0
SVDD	Pixel supply voltage	V	3.3
VDD	Positive supply voltage	V	2.5
VT	Treshold voltage	V	0.5 . . . 0.7
VOFFD	Dark offset voltage	V	0.09
VOFFB	Bright offset voltage	V	0.009
Vlog	Node voltage inside pixel	V	2.4 . . . 2.8

15.2 Abbreviations

3D	3 dimension
A/D-Converter	Analog-to-digital converter
ADC	Analog-to-digital converter
AOI	Area of interest, also named ROI (region of interest) or subframe
ASIC	Application specific integrated circuit
BGA	Ball grid array
BIOS	Basic input output system
BRDF	Bidirectional reflectance distribution function
BSSRDF	Bidirectional scattering-surface reflectance distribution function
Camera Link™	Interface for digital cameras
CCD	Charge-coupled device
CCTV	Closed-circuit television
CE	Communauté Européenne
CIE	Commission Internationale de l'Eclairage, International Commission on Illumination, Vienna
CMOS	Complementary metal oxide semiconductor
C-Mount	Mechanical lens mount standard
COB	Chip-on-board
CPU	Central Processing Unit
CRT	Cathode-Ray Tube
DSP	Digital signal processor
EM	Environment map
FLASH	Digital non-volatile memory
FPN	Fixed-pattern noise
GI	Gobal illumination
GPU	Graphics processing unit
HD	High definition
HDR	High dynamic range
HDRC	High dynamic range CMOS
HDRC-sensor	High dynamic range CMOS sensor
HDTV	High definition television
hil	Hema imaging library
hipe	Hema image processing environment
HVS	Human visual system
ICC	International color consortium
JND	Just noticeable difference
JPEG	Joint Photographic Experts Group (also image format)
LAN	Local area network
LCC	Leadless chip carrier

Index

Springer Series in

ADVANCED MICROELECTRONICS

1 **Cellular Neural Networks**
Chaos, Complexity and VLSI Processing
By G. Manganaro, P. Arena, and L. Fortuna

2 **Technology of Integrated Circuits**
By D. Widmann, H. Mader, and H. Friedrich

3 **Ferroelectric Memories**
By J.F. Scott

4 **Microwave Resonators and Filters for Wireless Communication**
Theory, Design and Application
By M. Makimoto and S. Yamashita

5 **VLSI Memory Chip Design**
By K. Itoh

6 **Smart Power ICs**
Technologies and Applications
Ed. by B. Murari, R. Bertotti, and G.A. Vignola

7 **Noise in Semiconductor Devices**
Modeling and Simulation
By F. Bonani and G. Ghione

8 **Logic Synthesis for Asynchronous Controllers and Interfaces**
By J. Cortadella, M. Kishinevsky, A. Kondratyev, L. Lavagno, and A. Yakovlev

9 **Low Dielectric Constant Materials for IC Applications**
Editors: P.S. Ho, J. Leu, W.W. Lee

10 **Lock-in Thermography**
Basics and Use for Functional Diagnostics of Electronic Components
By O. Breitenstein and M. Langenkamp

11 **High-Frequency Bipolar Transistors**
Physics, Modelling, Applications
By M. Reisch

12 **Current Sense Amplifiers**
for Embedded SRAM in High-Performance System-on-a-Chip Designs
By B. Wicht

13 **Silicon Optoelectronic Integrated Circuits**
By H. Zimmermann

14 **Integrated CMOS Circuits for Optical Communications**
By M. Ingels and M. Steyaert

15 **Gettering Defects in Semiconductors**
By V.A. Perevostchikov and V.D. Skoupov

16 **High Dielectric Constant Materials**
VLSI MOSFET Applications
Editors: H.R. Huff and D.C. Gilmer

17 **System-level Test and Validation of Hardware/Software Systems**
By M. Sonza Reorda, Z. Peng, and M. Violante

Printing: Krips bv, Meppel
Binding: Stürtz, Würzburg